T0205484

Library of Ethics and Applied Philosophy

Volume 38

The Library of Ethics and Applied Philosophy addresses a broad range of topical issues emerging in practical philosophy, such as ethics, social philosophy, political philosophy, philosophy of action, It focuses on the role of scientific research and emerging technologies, and combines case studies with conceptual and methodological debates. We are facing a global crisis, which raises a plethora of normative issues, but also poses a challenge to existing conceptual and methodological resources of academic philosophy. The series aims to contribute to a philosophical diagnostic of the present by exploring the impact of emerging techno-scientific developments on zeitgeist, collective self-image, and worldview. Concrete and urgent emerging issues will be addressed in depth and from a continental philosophical perspective, which may include dialectics, hermeneutics, phenomenology, post-phenomenology, psychoanalysis, critical theory and similar approaches.

More information about this series at http://www.springer.com/series/6230

Laurens Landeweerd

Time, Life & Memory

Bergson and Contemporary Science

 Springer

Laurens Landeweerd
Faculty of Science, Institute for Science in Society, Section of Philosophy
Radboud University Nijmegen
Nijmegen, The Netherlands

ISSN 1387-6678   ISSN 2215-0323   (electronic)
Library of Ethics and Applied Philosophy
ISBN 978-3-030-56855-9   ISBN 978-3-030-56853-5   (eBook)
https://doi.org/10.1007/978-3-030-56853-5

This Springer imprint is published by the registered company Springer Nature Switzerland AG
The registered company address is: Gewerbestrasse 11, 6330 Cham, Switzerland

Preface

The watchmaker's philosophy: each moment in time encompasses all events taking place in the universe, whilst each place in the universe can only contain one event at a time: a universe containing all places, and a time containing all events. This is the classical view on the nature of space and time, in which all events, past, present and future, of all things, in their places, are laid down as the parts of a watch on a watchmaker's table. But time itself can only be thought as process. From change, all derives, but change itself is not derived.
(Anonymous)

In spite of the relative obscurity of his oeuvre after the 1940s, Henri Bergson (1859–1941) was one of the most widely read philosophers of his era. He was the second child of seven and had a Polish father and Brittish mother. His father, Michal Bergson (1820–1898) was a composer; his mother, Katherine Levison (1834–1928) was the daughter of a Yorkshire doctor, and of English and Irish descent. Henri Bergson grew up in both France and England. In 1891 he married Louise Neuburger, a cousin of Marcel Proust. Proust's *In Search of Lost Time* (French: *À la recherche du temps perdu* (1913–1927)) was inspired by Bergson's ideas on memory.

The popularity of Bergson's ideas was in part caused by the position he took with regard to the tension between science, metaphysics and religion. The 'disenchantment of the worldview' (a term coined by sociologist Max Weber (1864–1920) to describe the effects of modern science on society) met with quite some resistance around the turn of the century. The European public was frantically seeking for answers to questions of the soul and the nature of life. The impact of Darwin's *Origin of Species* was immense, and led to a polarisation regarding topics such as the origin of life, evolution theory, religion, the ideals of scientific positivism and the emergence of spiritualism: on the one hand, there was public enthusiasm concerning the promises of scientific knowledge and technological progress, on the other hand, many were critical of the dehumanising effects of these developments. The historicist tradition in philosophy that was established by Georg Wilhelm Friedrich Hegel (1770–1831) informed an active opposition to these developments. At the end of the century this opposition was supported by thinkers such as Wilhelm Dilthey (1833–1911) and Friedrich Nietzsche (1844–1900) – specifically because of his endorsement of irrationalism. Fitting in that historical niche, Bergson's writings drew much attention. In the midst of his lifetime Bergson was held in high regard, not only as a philosopher but also as a public celebrity. Not only his philosophical writings resounded with the Zeitgeist of the 1910s, he also received broad acclaim as a political figure. Bergson played an important role in 1917, when French Prime Minister Aristide Briand, being aware of the philosopher's reputation of self-critical integrity, asked Bergson to try and convince United States President Woodrow Wilson to join the war effort. And after the war, Bergson was asked to negotiate between the different parties that needed to broker peace; his views on arms control were ahead of his time and he preferred diplomatic solutions over war, although some (e.g. Martin Heidegger) perceived of his diplomatic role after the First World War as flawed.

Bergson was a Frenchman of Polish-Ashkenazi and English-Irish-Ashkenazi descent who spent his childhood in London. In an age when nation-states had come to be seen as natural entities, this background lent him an aura of supranational neutrality. In 1922, he was appointed president of the International Commission for Intellectual Cooperation, an organisation that was the precursor to UNESCO. Bergson won the Nobel Prize for literature in 1927 in recognition of 'his rich and vitalising ideas and the brilliant skill with which they have been presented'. And although this might lead some to believe that his work should be identified with literature and poetry, he was also acutely aware of the state of the art in the sciences of his age.

Bergson's oeuvre spanned a wide range of fields and topics, including the most basic questions in philosophy and science: the nature of life, the nature of time and the nature of consciousness. But in spite of the strong involvement of his philosophy with debates concerning the sciences of his age, he was largely forgotten as a philosopher of science after his death. An important reason for this collective amnesia lies in the reception of the debate between him and Albert Einstein.

On 4 April 1922 the *Société française de philosophie* organised a meeting of minds between Einstein and Bergson. The ensuing debate counts as a turning point in the relationship between the exact sciences and the humanities. Bergson's views on time seemed to conflict not only with the concept of time as it existed in classical physics but also with the views on time Einstein had elaborated in his theories of relativity. Einstein's theory of general relativity became widely accepted and as a result, Bergson's views on physics have mostly been regarded as flawed, in scientific and philosophical circles alike. But whether Bergson indeed did not accept the basic strands of Einstein's scientific views can be contested: whilst Bergson discussed Einstein's theory elaborately, Einstein only responded with a brief commentary on Bergson's point of view, stating that the time of the philosophers did not exist (Canales 2016, pp. 5). To his mind, there was only the (objective) time of physics, and besides that the subjective experience of time, which he regarded as merely psychological in nature. After the debate, Bergson wrote *Duration and Simultaneity* (French: 'Duration et simultaneité' (1922)) to elucidate his viewpoints on time and on Einstein's relativity. In the years after 1922 the consensus view was that Bergson lacked sufficient understanding of the basic mathematical principles underlying the general theory of relativity, and more generally, that philosophy should be regarded as irrelevant for exact scientific thought. This is confirmed by the authoritative study on the debate by Jimena Canales published in 2016, *The Physicist and the Philosopher* (Canales 2016) who defends the aforementioned view that a main source for the crisis between the exact sciences and metaphysics lies in the schism that arose between Henri Bergson and Albert Einstein[1]. Their theoretical opposition was to evolve into a schism between the exact sciences in general and 'continental philosophy'.

The term 'continental philosophy' is a problematic misnomer. It currently describes an amalgam of rather unrelated philosophical traditions, varying from France to Germany, presenting the divide analytic/continental as a universal fact in philosophy. The term 'continental' was pejorative. It stems from the British reception of German philosophy (Mullarkey 2010), specifically from early Victorian England's reception of Hegelian dialectics (Vrahimis 2011) and Schelling's idealist philosophy. The poet Samuel Taylor Coleridge's was inspired by German Idealism (Schelling and Hegel's works in specific). But adherents to Jeremy Bentham's ideas disliked such foreign influences (Vrahimis 2011). Coleridge's 'exotic adherence' to Hegel was dismissed by them as a continental sympathy. John Stuart Mill was the first one to use the term 'continental philosophy' to refer to the Colridgeans. In spite of Immanuel Kant's Prussian origins, the Benthamites did adopt his ideas, whilst Schelling's German Idealism and Hegel's dialectics came to be seen as a problematically exotic tradition[2]. The world might have forgotten about this divide, if it were not for Bertrand Russell's dislike of Bergson's philosophy, and now, most European and American students in philosophy take the notions of 'analytic' and 'continental' as universally given historic traditions. In recent years, the term 'continental' came to be appropriated by French and German philosophers as nom de guerre to create a 'reverse discourse' (Foucault 1976). In this book, I prefer avoiding both terms as much as possible.

The crisis between philosophy and the exact sciences is broader than metaphysics alone: it is consistent with the breach between the exact sciences and the humanities – including sociology, ethnography, anthropology, history, etc. And as a result, philosophy is no longer seen as relevant for contemporary science. This breach between two cultures was also observed by C. P. Snow in the late 1950s (Snow 2001 [1959]), and the gulf between these two populations continues to grow. The problem was also observed in by Michel Serres (Serres 1980), who, in his 'Hermes V' likened the bridging between the exact sciences and the humanities to navigating the Northwest Passage (discussed with Latour (Serres and Latour 1995, p. 70; Yussof 2011)). And in the introduction to an edition of Serres' works Harari and Bell state: '[…] we ordinarily conceive of two populations: the scientists without culture (educated but not "cultivated") and the humanists without scientific knowledge (cultivated but not "educated")' (Harari and Bell 1982, pp. xii). Like the notorious Franklin expedition of 1845 that in its quest for such a passage got

---

[1] Other reasons contributed to the relative lack of attention for his philosophy after the war, in specific his views on the phenomenon of life (see Chap. 3).

[2] Seen the recent political upheaval over the Brexit, not much seems to have changes in the United Kingdom.

stranded on Beechey Island at the west coast of Greenland, Bergson's influence got stranded somewhere between the sciences and the humanities. Navigational systems of the time carried along an overly confident hope for a passage, but in the end, the conceptual instruments built for a meeting of minds were too different to break the ice.

Philosophers have a tendency to discuss science in much too general terms whilst often referring to research fields and scientific discussions from the past as frame of reference, approaching science from an outsider's perspective. For the exact sciences themselves, such generalist philosophical ponderings became increasingly irrelevant to their own field and research practice. This however does not mean that both areas of study are doomed to a definitive separation. The climate is changing, and seas that could not be sailed in the past may now be sailed at last. Both cultures of thought may mutually benefit from a rekindling of the dialogue. But in order to do so, it is necessary to map the Northwest Passage. To mend the breach between these two academic cultures, one needs to revisit the moment of divorce and assess the nature of the separation.

An undertow between the two streams of thought did persist, interestingly through the aforementioned 'continental' tradition. The crisis did not destroy all lines of communication: several thinkers have focused on the content (rather than the methodology) of the exact sciences. In France, several thinkers did follow suit on Bergson's thought and although some focused on philosophy itself (Gilles Deleuze, Maurice Merleau-Ponty), others continued the discussion of science. These include, apart from the above-mentioned Michel Serres, amongst others, Pierre Teilhard de Chardin, George Canguilhem, Gilbert Simondon and Bruno Latour. Bergson's philosophical influence on the international scientific debate, however, remained relatively marginal – although some scientists did refer to his work in the years after the Second World War, including the quantum physicist Louis de Broglie and the psychologist Jean Piaget (albeit grudgingly). These small passageways, opened shortly through the ice, do not suffice for a true communication.

The early 1900s formed a battleground between two opposing worldviews: mechanistic determinism and organic process thought. To be able to understand how these concepts of time affect scientific discourse, both in the past and in contemporary science, it is necessary to have a closer look at them. I will do so through a discussion of neo-Kantian philosophy, the criticism raised against this school by phenomenology and the distinction between intellect and intuition as it is discussed by Bergson.

The exact sciences tend to approach their object from without, from the point of view of an external disinterested detached observer. And specifically in Bergson's era, mechanistic tendencies that characterised the exact sciences from the seventeenth century onwards were still a dominant perspective – to an extent, they still are. But alternatives were beginning to become more and more apparent. At the heart of these debates was the opposition between two worldviews: the mechanistic worldview as it developed during the scientific revolution and process thinking.

Bergson was one of the most important contributors to the debate between these two views. In both, differing concepts of time existed.

The stake between mechanistic thought and process thought was high: the debate concerned the way in which we would need to understand the nature of the universe. Whilst many physicists around the turn of the century felt a need to reaffirm positivist thinking against the tendency of the Zeitgeist towards what they regarded as mysticism, some saw a continuation of the deterministic principles of Newton and other great minds of the era of the scientific revolution as problematic. To them, the merger of rationalist mathematic and geometric principles as the basis for understanding life, the psyche and the universe did not provide viable answers but rather hindered the acknowledgement of human flourishing and the complex unpredictability of life. They were critical of the concept of deterministic causality as it came to be articulate in Newtonian mechanics[3], and as it was defined by someone like Julien Offray de La Mettrie and later authors such as Pierre-Simon Laplace. And although this view on causality was still regarded by many as the 'only type of rational order' (Čapek 1971, p. 292), alternative ways of viewing the world were being explored by scientists and philosophers alike.

Scientific debates around the turn of the century were deeply intertwined with the fin de siècle mentality. Society was concerned over the impacts science was having on existing views on the origin of life, the existence of the soul and the nature of the universe. It saw these as threatened by the insights that were discussed within biology, medicine, psychology, chemistry and physics. Scientific progress radically affected the implicit metaphysical, theological and political views current at the time. As one of the major thinkers about time, Bergson represented a post-deterministic view and elaborated a philosophy that, as I will argue, still holds relevance to the sciences.

Bergson's *Creative Evolution* (L'Évolution créatrice 1907) attempted a fundamental redefinition of philosophy. Philosophy, for Bergson, needed to think beyond the human condition. It needed to go beyond the frame of our evolved and dominant habits of representation. Specifically, our ways to represent everything in spatial terms needed to be challenged (Caeymeaux 2013).

Bergson believed that we tend to confuse duration with extension. The spatial metaphor for our view of the world, as determined by our focus on the sense of vision and underscored by the sense of touch, echoes Bergson's criticism of the epistemology of Immanuel Kant. Bergson saw Kantian epistemology as problematic since it tempts us to think time in terms of space. Kant considered 'reality' as a problematic concept for our knowledge systems. Whilst attempting a reconciliation between rationalism and empiricism, his focus was on the formal conditions of knowledge. But in this focus, Kant could not escape referring to something beyond perception (the thing upon itself), as opposed to a knowing subject on the other. Our senses, our formal cognitive categories of space and time, etc. are, within this epistemology, the aspects that build up our consistent knowledge of the world. Although

---

[3] Although Newton also wrote about astrology, something his later heirs had been keen to obscure.

Kant in part provided an alternative to Descartes' dualism, his philosophy remains dualistic. This is due to his definition of the limiting boundaries of a formally stated 'knowledge subject' and this subject's limited perspective on 'what might be known' (the phenomenal world) versus the noumenal (the things in themselves). This led to a further problem in the theory of knowledge.

Many neo-Kantian philosophers of the nineteenth and twentieth century unfortunately tended to equate the Kantian knowing subject (the formal instance of knowledge) with a 'self' in the psychological sense of the term. This led them to the idea that knowledge might be determined as an instance of the mind, a mere psychological entity. This 'psychologistic' interpretation neo-Kantian philosophy led Edmund Husserl to separate contents of knowledge from the areas that this knowledge refers to,[4] which gave rise to the phenomenological tradition. Bergson's position on this issue was that he embraced the idea that only immediate experience of consciousness could be taken at face value.

Neo-Kantian philosophy, in its application on the human mind, takes a strongly materialist position. It takes the mind to function as a mechanism. In this regard it is a direct successor of the mechanistic worldview. The mechanistic worldview is the metaparadigm that informs an interpretation of phenomena in the world as if they were a clock. Theories built on the basis of this worldview reflect the idea that our thoughts and actions are causally determined by underlying material causes and are ultimately fully predictable. They are phenomena governed by the same laws as those that apply within Newtonian mechanics. We should therefore be able to understand their nature and emergence through systematic, empirical research methods. The externality of causation is also taken as a given. It is either an external cause (or, to put it in Aristotelian terms, an 'efficient cause') or an internal cause, residing in our biochemical and ultimately physical makeup (in Aristotelian terms, 'material causes'). In practice, these two causes can still be recognised in modern psychology: environmental causes (education, society, etc.) and physical causes (hormones, neurons, etc.). Phenomenology, as critical countertradition, regarded such reductionism as flawed.

A well-known element of phenomenology is the moment of *époche* in the phenomenological reduction: Husserl chose to put the notion of 'reality' between inverted commas, focusing the attention on phenomena as they appeared to us rather than retaining an accustomed notion of an accessible reality. He saw that the only strategy that would enable a discussion of knowledge required an abolition of the 'order of the real', since it already formed part of our way of perceiving the world rather than that world itself.

Here, Bergsonian philosophy has an important thing to contribute: Husserl's notion of 'putting reality between inverted commas' still creates a separation

---

[4]This issue would be discussed by other philosophers in reference to the concept of perspectivity. It would also be taken up by Ludwig Wittgenstein, in his famous comparison between the eye and the visual field (Wittgenstein 1921) in his *Tractatus Logico Philosophico*. Wittgenstein would phrase the issue as a faulty reduction of the inner perspective to a phenomenon appearing in the outer perspective.

between thought and reality, bending the human self back onto itself – even if only to acknowledge that reality 'as it exists in itself' can never be known[5]. But the Kantian 'thing in itself' haunted phenomenological philosophy as well[6]. Although, things as they are supposed to exist outside of perception, awareness or knowledge remain forever beyond our experiential horizon. As such, the question of our belief in it is irrelevant. It is at this point that Husserlian phenomenology tacitly acquiesces with Kant[7]. Bergson however diverges from Kantian philosophy at this point. In Kantian philosophy, the knowing subject[8] is presupposed as a condition for all intuitions. The noumenal realm thus remains forever beyond the experiential horizon of that subject. In *Matter and Memory*, Bergson (1988 [1896], p. 75) states that the need felt within Kantian philosophy to combine the extremes of idealism and realism falls away when one starts from intuition, rather than from a formal knowledge subject. He thus considers the intangibility of the world as it exists upon itself as a result of an overly narrow interpretation of knowledge in Kantian philosophy: Kant, to his mind, restricted our ways of knowing the world too much to how it is constituted through our forms of sensibility of time and space. Although he admits that this might be the case for positivist science, it is not the only source of knowledge we have. Intuition may open up the doors to the ineffable where the analytic method cannot (Bergson 1999 [1903]).

Some scholars, such as Alexis Philonenko (1994) or Camille Ricquier (2016), consider Bergson's views as a reconstruction of Kantian epistemology that takes the Cartesian notion of intuition more aux sérieux: '[…] Bergson finds in [Kant] a powerful ally for penetrating farther into the metaphysical deepening of intuitive reality, into which Descartes had already begun to enter. The Cartesian intuitions—weak, marginal, and soon rejected by the Cartesians themselves—will find in the Critique of Pure Reason one of the instruments for their intensification. Kant had taken a step backward with respect to Descartes' advances. Nevertheless, Bergson's Kantianism will take a step forward and will be—in his own words, which must be taken à la lettre—a "revivified Cartesianism"' (Ricquier 2016). This revised interpretation carries along that Bergson's philosophy should not per se be read as an opposite of Kant's philosophy, but rather as a strengthened revision thereof.

There are parallels between the works of Bergon and phenomenology. And indeed, when asked about phenomenology's stance to Bergsonian philosophers, Husserl is said to have replied 'We are the true Bergsonians' (Hering 1939). One might argue that Husserl's phenomenology is complementary to Bergson's oeuvre, in its reflection on a specific form of subjective experience: the Cartesian ego, the subject of science, whereas Bergson takes a step backward in returning to intuition as the primal way of being-in-the-world.

---

[5] Although we are used to think of the world as if it were external to us, and as if we were external to it, this world as we think it is always by necessity a world that already contains us thinking it.

[6] Belief in it, in phenomenology, is suspended, rather than either affirmed or denied.

[7] Although for some Husserlians this probably amounts to blasphemy.

[8] This formal constitution of a knowledge subject should not be confused with our concept of 'self': this is the mistake made by neo-Kantian philosophy which would be refuted by Husserl.

In *Time and Free Will*, Bergson discusses two apparently conflicting aspects of reality: multiplicity and continuity. How can the world be both a collection of individual things and something continuous? This is only possible when regarded from a temporal perspective. One cannot grasp continuity on the basis of a geometrical conception of time. It is for this reason that concepts of time that define time as a 'fourth' dimension ultimately remain unsatisfactory (see Chap. 3).

The multiplicity of the world is continuously given in an immediate sense to consciousness. 'World' and 'subject' should be regarded as two derivate aspects of this immediacy. This brings a nuance to the epistemological position of phenomenology. The dilemma here lies in the phenomenological stance to put 'reality' between inverted commas (the method of *epoché* first suggested by Husserl in the second edition of the *Logical Investigations* (*Logische Untersuchungen*, Husserl 1900 [1911])). It may be regarded as a radicalisation of the methodological constraint, already to be found in the *Logical Investigations*, often without epistemologically neutralising the metaphysical consequence of the term reality.

With phenomenology, Husserl adopted a methodological philosophical approach to experience. But committing to methodology sacrifices something important: Husserl was aware of the dilemma, seeing that the formal instances of subject and object are not defined a priori, but given a posteriori. They are not given in direct experience, but presupposed in the conceptions we employ to structure our experience. For Husserl however, the 'prima realitas', or first basis for any metaphysical thought, is the intersubjective space, which enables our view of 'me as me', and our view of 'object as object'[9]. It therefore is social and discursive from the outset – albeit in a formal sense.

The original intuition of phenomenology was that our experience should be regarded as more valid as first point of departure than rational argumentative structures. In short, the phenomenological approach is a means to orient the subject onto its own experiences. Any structure of reasons and arguments does not emerge out of nothing, it can only be based on experience. But the phenomenological method consists of its notorious 'bracketing' of any notion of the world 'outside of experience', a devotion to a functional philosophy of, ultimately, a pragmatic nature: its methodology turns to a focus on how the subject experiences, not on how experience occurs before subjectivity. Husserl himself was well aware of this, and alluded to the restrictions of his approach in phenomenology on numerous occasions. But the edification of this subjectivity – as intersubjectively, and thus ultimately discursively constituted – already leads to a dismissal of what Bergson points to in terms of duration, vital impulse and immediate concreteness. For Bergson the immediate should be retained as prima realitas. The immediate data of consciousness are pre-discursive, and not given by an intersubjective space, nor given through convention-led discourse between people, but in sensory experience, emerging before any individuality is unveiled, be it the self or the individuality of its object.

---

[9] 'Etwas als Etwas', or knowing something *as* something was already recognised as a modality of cognitive determination by Immanuel Kant.

Mechanistic thought by its very nature uses the intellect to analyse the world in terms separated elements – the distinction between subjective experience and an objective world as the most basic of these. But from intuition, the organic and fluid nature of the world is allowed to surface. The distinction between the analytic mind that moves around the object and intuition that enters into it, relates to a distinction that Bergson describes in earlier works: Bergson, in his distrust of the philosophical debate of the distinction between concept and world (word and thing, symbol and event), aims to avoid the problem altogether, by adhering to 'intuition' of the nature of things/events as a more direct source of knowledge, which can move beyond the restrictions of our linguistically determined abilities of describing and explaining (Čapek 1978). But in terms of our knowledge systems, representing such experience always poses a problem.

In *An Introduction to Metaphysics* (1903 [1912]), Bergson gives an account of the difficult relation between representation and object, stating: 'A representation taken from a certain point of view, a translation made with certain symbols, will always remain imperfect in comparison with the object of which a view has been taken, or which the symbols seek to express. But the absolute, which is the object and not its representation, the original and not its translation, is perfect, by being perfectly what it is' (Bergson 1903 [1912]). Here, the sciences are caught in a difficult position, specifically if we take the exact scientific approach to reality to be predominantly analytic and intellectual rather than intuitive and introspective. Leaving chaos theory and complexity theory aside, most approaches in the exact sciences still remain analytic and reductionist: it is only by understanding the parts and their relation to each other that one understands the whole. Scientists analyse by interpreting, translating, using symbols and representations in the shape of words, formula and imagery. In all these cases, what Bergson refers to as the object, which to him is 'absolute', remains beyond the horizon of these traditional analytic knowledge strategies. To be able to have access to the absolute object, we should stop moving around it, but rather enter into it (Bergson 1903 [1912]). Here, the experience of the object is of a different nature. But the only 'object' or reality that we can *truly* enter and seize from within is our self. And this, we can only experience in immediate consciousness.

Questions about the origin of life, the nature of time and the role of memory remain key questions for many scientific fields. At the same time, the current age encompasses other priorities. Climate change, tinkering with the basic fabrics of biological systems – including our own nature – is high on the agenda. They are closely interrelated with discussions about political and industrial agendas, public responsibility in innovation and discussions about the nature of progress. But to be able to understand what is steering the debate, it is necessary to render explicit the basic positions that inform it.

Although some interest in Bergson's ideas persisted, it mostly remained a marginal strand of scholarly and scientific discourse. The following chapters seek to revisit the challenges Bergson laid down in his work for philosophers and scientists alike. In this regard, I will address, as said, three areas in the natural sciences: physics, the life sciences and neuroscience. Then I will discuss the relevance of Bergson's ideas for contemporary converging sciences and technology. In the upcoming three chapters, I will follow a similar design: I will first outline the context of the discussion of the discipline under consideration at the time of Bergson. Subsequently, I will outline Bergson's specific views in this regard more in detail. And finally, I will show how the discussions at that time are still relevant for the discipline in question in the present. Bergson's philosophy will thus be discussed within the context of his debate with various scientists and philosophers in his day and age, revitalising these debates by extrapolating them to current and ongoing discussions in three domains of scientific research: physics, the life sciences and neuroscience. The aim of this book is an elucidation of Bergson's core ideas, but not in the sense of author studies. The focus remains on the relevance of these ideas for contemporary science.

To investigate whether Bergson's ideas hold any relevance for modern science means sailing through dangerous waters. It means readdressing the deeply problematic debate between Einstein and Bergson, it means addressing sensitive discussions concerning definitions of life in the light of vitalist theories, and it means addressing the persistent issue of the mind-body dualism in neuroscience. Only by pursuing these issues can the basis of the fissure between these two areas of knowledge be uncovered. The objective of this publication, in part, is to bridge this very fissure.

Bergson's views on science have specifically been discredited as a result of his debate on time with Einstein – a result of an unfortunate misreading by both physicists and philosophers. French philosophy after the Second World War turned to the works of Edmund Husserl, a contemporary of Bergson, and his pupil Martin Heidegger, rather than to the oeuvre of Bergson, although he influenced both. And besides the unfortunate debate with Einstein (after which Bergson's position came to be misrepresented), Bergson became associated with nineteenth-century vitalism; crudely explained: the view that there is a mysterious force or vital spark at work in living beings, making them different from non-life (stones, coffee mugs, stars, dead wood, asteroids, crystals, gases, temples, photons, cars, etc.). This was, again, a misrepresentation of Bergson's actual position[10], although Bergson was indeed also interested in more spiritual subjects, including religion and Catholicism. Thus, another reason for the collective amnesia after Bergson's death was that younger generations were less inclined to the mystical, endorsing a modernist functionalistic view of science and technology. As a result, Husserl's phenomenology, that aimed to be a rigorous science, seemed preferable over Bergson's philosophical oeuvre. Advances in genetics during the 1930s seemed to fill the gap that was at the time still open with regard to the mechanisms behind evolution: any form of vitalism, be it that of the biological vitalists of the nineteenth century or the philosophical concept of vital impulse as it was expressed by Bergson, were regarded as out of date. A further contribution to the collective amnesia mentioned above was that Bergson's oeuvre was perceived to be too unsystematic and anti-methodological. Bergson himself did not always help his cause either: he asked his wife to destroy

---

[10] Which might be termed 'panvitalist', thus referring to Bergson's view that vital impulse is not merely a distinguishing characteristic of biological life forms (nor is duration).

his archives after his death, so that no unfinished manuscripts survived. Furthermore, given the complexity of Bergson's philosophy, he has hardly been 'canonised' in any existing tradition (Mullarkey 2010).

A rehabilitation of Bergson is long due, and has only been spurred recently, by (a.o.) the aforementioned study by Jimena Canales. As Canales describes it: 'Einstein searched for consistency and simplicity, Bergson focused on inconsistencies and complexities' (Canales 2016, pp. 21). This led to a clash between on the one hand the static tradition Einstein adhered to, claiming that science could reveal the universal and eternally valid laws that determined the universe, and on the other hand the dynamic tradition Bergson adhered to, claiming that never-ending change was the basic aspect of reality[11]. Bergson's ideas on time are not regarded as obsolete by all scientists. They are still deemed to be relevant by several proponents of quantum theory, the principles of which clashed with the determinist position embraced by Einstein. This book aims to continue the effort of those involved in rehabilitating the works of Bergson, albeit not without critical assessment. It also aims to contribute to a wider awareness over the interrelation between physics, the life sciences, the neurosciences and philosophy.

Rather than being devoted to incremental, piecemeal research, current developments in physics, biology and neuroscience indicate that science is once again committed to addressing the basic questions, the ones it shares with philosophy: definitions of life, the place of consciousness and the nature of time. This provides an opportunity to reopen the traffic over these bridges and revitalise a philosophy of science that is prepared to discuss content rather than focusing either on the logic of scientific method or on the sociology of science. By reconsidering Bergson's contribution, we pick up the debate precisely from where it became stranded a century or so ago. This book attempts to reconnect the sciences with the humanities by both a clarification of Bergson's ideas on time and a revitalisation of his views for contemporary science.

In the following chapters, I will discuss how the (apparently quite diverging) core motifs of Bergson's philosophy (time, life, consciousness, experience, morality, etc.) relate to each other and how they are relevant for understanding the Zeitgeist of our own time. I will focus on several of Bergson's publications[12]: *Time and Free Will: An Essay on the Immediate Data of Consciousness* (French: 'Essai sur les données immédiates de la conscience' (1889)), *Matter and Memory* (French: 'Matière et Mémoire' (1896)), *Introduction to Metaphysics* (French: 'Introduction à la métaphysique' (1903)), *Creative Evolution* (French: 'l'Évololution créatrice' (1907)), *Duration and Simultaneity* (French: 'Durée et simultanéité' (1922)) and *The Two Sources of Morality and Religion* (French: 'Les deux sources de la morale et de la religion' (1932)). Through these, I will discuss Bergson's dialogue with and position in the debates concerning the sciences of his day and age, but I will also

---

[11] In fact an ancient opposition that already existed in presocratic philosophy, as I will discuss later on.

[12] Here I will mention only the year of the first French publication. Elsewhere I will use English translations (except where French is needed), mentioning the year of the specific edition I used.

point out how the undertow of his thoughts resurged in various later scientific developments, thereby indicating how his thoughts are highly relevant for contemporary issues related to the convergence of science and technology.

In its effort, this book also hopes to contribute to a general strengthening of the role of philosophy for the sciences. When any scientific field is confronted with problems related to the basic presuppositions of its discipline, this inevitably entails a shift into a different mode of thinking, namely philosophical reflection (Zwart 2017). Scientists' interest in philosophical debate often only occurs when something is found to be inconsistent within the existing dominant paradigm. On such occasions, scientists tend to opt for a more public outlet of their concerns and ideas. Unfortunately, when data apparently correct foundational philosophical concerns, the hatches of the disciplinary windows close again. Afterwards, those scientists that do continue to venture into philosophical debates are usually met with scepticism by their colleagues. Since this holds an important problem for the critical mass of scientific praxis, the dialogue between science and metaphysics is in urgent need of restoration. This book therefore aims to support a reflexive turn in science, as well as a turn towards contemporary science in philosophy.

In the following chapters, I address three fields of science: physics in relation to the abstract and the concrete, biology (notably emerging research trajectories in synthetic biology) in relation to concepts of life and the neurosciences in relation to the technical nature of human identity. The discussion in these domains revolves around one main topic: time. Time, isolated from experience as the measure of events in the universe in modern physics; time as the measure of emergent systems in evolution and as the backdrop of the theory of evolution in biology; time in relation to memory and imagination in neuropsychological accounts of memory. In these treatments, I will discuss the ideas of Henri Bergson as a basis to unveil time as a living process, rather than as a mechanism for the recording of events.

Bergson is not a philosopher of science par excellence. Conventionally, the philosophy of science is either about defining norms for epistemic validity or the logic of scientific rationality. But where the conventional philosophy of science deals with issues of knowledge validity and methodological legitimacy, Bergson's contribution to the sciences lies in his interpretation of the metaphysical undertow of scientific discourse. Bergson considered scientific theories and practices as moments in an evolving historic processes. As such, they need to be discussed from a historical perspective, but not reduced to those processes. He thus engaged in a dialogue with the sciences, critically challenging and probing their basic claims.

For this book, I combined my experience of multiple national and international projects in ethics and policy of science and technology with my parallel research in philosophy. They represent a contextual reading and interpretation of the oeuvre of Henri Bergson. I departed from a reconnaissance of literature within these three domains of science that referred to Bergson, which became the basis of a reconnaissance of the oeuvre of Bergson. On this basis I sought to distil a message from this combined reading for contemporary science, and on that basis to diagnose the role and place of technology in this day and age. Methodologically speaking, the approach to this book thus amounts to a reading of Bergson's oeuvre in a contextual relation to the three domains discussed. It travels through a tripartite move from theoretical physics through the life sciences to the neurosciences. These steps are necessary to discuss the creative coevolution of science, technology and biology in the ensuing chapters. The core question that lies behind this endeavour might be circumscribed as: 'how can change be thought in the context of science and technology?' The more pragmatic question of this endeavour might be: 'How can a contextual reading of the works of Henri Bergson contribute to a better self-understanding in the sciences – for first three domains, then the relation of these domains to technology?'

This book does not seek to merely add another volume to the philosophical library, but rather to revitalise this particular branch of the story of European thinking. At this turning point in history, humankind needs to take on an unprecedented responsibility: we need to grow beyond the spoilt luxury of being in a world that nurtures us and become aware of the need to nurture a world, both in ecological and economic terms. Understanding the stakes of the worldview of that era is of the utmost importance for the future prospects of our own species. This age of ours, this 'now' in which we live, in which the future has become present, is the age in which the demise of humanism should indeed be mourned, and our inability to replace it should be confronted. The sciences may play a crucial role in this challenge, but they cannot if they are not sufficiently aware of their own predispositions.

The ensuing chapters were written not only in dialogue with Bergson's texts, but also with a view to the human condition in the current age of technoscience. They build on publications on Bergson by others (author studies such as Canales's aforementioned *The Physicist and the Philosopher*, but also older studies such as Gilles Deleuze's 1966 study *Bergsonism* or Milič Čapek's 1971 study *Bergson and Modern Physics*). It also extrapolates from lesser known studies (international obscurity is at times due to non-availability in English) such as Jan Bor's 1992 study *Bergson en de onmiddellijke ervaring* 'Bergson and immediate experience') and Hein van Dongen's 2014 study *Bergson*. Building on the groundwork of these authors, I aim to support a venture into questions that take us beyond the laboratory and beyond state-of-the-art research. For ultimately, there is no competition between science and metaphysics. Whilst they may inform each other, they are distinct knowledge fields with distinct approaches and distinct purposes. These distinctions should be drawn sharply, so that both might benefit from the findings and insights from both sides[13]. Although the reflections involved remain essentially at the heart of experimental thought, they ask the reader to question the foundations of the positions and theoretical perspectives usually accepted as dominant in various science areas. It supports critical self-investigation in current scientific practices, not merely to awaken the slumbering philosophical impetus of various scientific disciplines, but specifically to trigger the willingness of scientists to reflect on the wider implications and cultural relevance of their findings and show how these are important to

---

[13] Not doing so raises discursive problems: for example, when contemporary neuroscientists feel tempted to present their findings as proof for the non-existence of free will, a confusion emerges between science and metaphysics.

contemporary society. The aim is to open up some of the 'disciplinary hatches' of scientific thought. Some may have been sealed for decades, others have remained open more or less, but only for those prepared to read publications of authorities in the field that go beyond the state of the art. Such reflection is necessary, since, as Bergson also held, when the hatches close, metaphysical presuppositions that constitute the point of departure for a praxis of scientific research are mistaken for their conclusions[14].

Nijmegen, The Netherlands   Laurens Landeweerd

[14] To illustrate this with an apocryphical story to which I found no reference beyond my father: once, during a psychology conference on behaviourist psychology – the school that considers man as behaving things without intentionality or free will – the behaviourist psychologist B. F. Skinner fell asleep during a lecture that dealt with his work. The speaker, unaware, delegated a question from the audience to Skinner, who, startled, answered with elegant academic aptitude that he couldn't answer the question since he could only see 'things behaving'.

Contents

Chapter 1
Introduction

> *Berkeley was unable to account for the success of physics, and,*
> *whereas Descartes had set up the mathematical relations*
> *between phenomena as their very essence, he was obliged to*
> *regard the mathematical order of the universe as a mere*
> *accident. So the Kantian criticism became necessary, to show*
> *the reason of this mathematical order and to give back to our*
> *physics a solid foundation – a task in which, however, it*
> *succeeded only by limiting the range and value of our senses*
> *and of our understanding. The criticism of Kant, on this point at*
> *least, would have been unnecessary […] if philosophy had been*
> *content to leave matter half way between the place to which*
> *Descartes had driven it and that to which Berkeley drew it*
> *back – to leave it, in fact, where it is seen by common sense.*
>
> *- Bergson, Matter and Memory, Introduction to the 1911*
> *English translation.*

A discussion of the relevance of Bergson's ideas for a philosophy of science cannot avoid controversy. There have been many critics of Bergson's work, specifically of his views on developments of the sciences of his day. These include philosophers (such as Bertrand Russell and G.E. Moore, but also more recently Reichenbach 1956) as well as scientists (most prominently Albert Einstein). As I remarked in the foreword, his ideas have supposedly contributed to the schism between the exact sciences and the humanities (Canales 2016), the latter apparently having lost the debate on the nature of time. Yet, the questions that he posed remain as relevant now as they were then: they apply to the debate on human identity in terms of neuroscience, to the discussion of the arrow of time in relationship to determinism and indeterminism in physics, and the question of the origin of life in biology. But they are also relevant for understanding the nature of scientific thought. A 'fringe

© Springer Nature Switzerland AG 2021
L. Landeweerd, *Time, Life & Memory*, Library of Ethics and Applied
Philosophy 38, https://doi.org/10.1007/978-3-030-56853-5_1

community' within the exact sciences continued to read and adhere to his ideas, many of whom remain authorities within their field.[1]

Bergson's criticism was specifically aimed at mechanistic thinking. Here, he did not merely criticise some of the scientists of his age, but rather a wide tradition in western thought that includes the ideas of René Descartes, Isaac Newton and Immanuel Kant. His critique concerned mechanistic thought in its deepest nature: a critique of our conditioned and habitual ways of conceiving of the world. Although he was a popular author in his own days, acclaim from the side of the exact sciences was scarce. He wrote in an age in which scientific completeness, radical analysis and universal validity clashed with subjective experience, spiritual discovery and the flow of life. Support from the exact sciences for someone who took a midway position between these two was not self-evident. Still, even after interest in his works began to wane, his ideas inspired theorists in quantum theory, physical chemistry, chaos theory and systems thinking. In most theories of science, Bergson's oeuvre is ignored, however. This study seeks to sensitise the reader to Bergson's ideas. It aims to show that his views represent not only an important undertow in twentieth-century philosophy, but also in twentieth-century science.

We should not ignore the intrinsic congruence between the exact sciences and philosophy. Science and philosophy share a similar basic drive:[2] to investigate the nature of things. But although science and philosophy seem to address similar questions (what is life? what is consciousness? what is time? etc.) they ultimately walk different paths to address these questions. The exact sciences have committed themselves to develop specific high-precision tools and methodologies. These enable an enhancement of knowledge through standardised empirical observation, organised and described through the use of mathematical tools and instruments. Philosophy positions itself in the area that precedes the development of such tools and instruments: it is in the first place guided by the effort to ask the most fundamental questions possible. Still, both science and philosophy seek knowledge. And both have an empirical dimension. But philosophical empiricism starts from introspective experience, whereas scientific research is based on experimentation. Neither is able to fully express the manifold of phenomena we encounter in the world. This shared inability is interconnected with the problem of representation: both science and philosophy are expressed in symbols that inevitably create taxonomies of the world. They compartmentalise, generalise and create orders of things, rather than referring to their concreteness and individuality. In other words, by naming the world we lose the ability to reach out to singular things. Bergson circumvented this by developing a specific approach to the nature of experience, conceptual representation and reflection.

---

[1] E.g. Louis de Broglie, Edgar Morrin, Ilya Prigogine.

[2] And there were times and places where the distinction between the two was not made: the title PhD, an abbreviation that is currently accepted worldwide as the title purportedly bestowing a mastership in any scientific discipline except engineering (or, in some cases and contexts, law), is a relic of this synergy: Philosophiae Doctor.

Bergson's earlier ideas were indebted to the philosophy of his mentor Étienne Boutroux (1845–1921), a pupil of the physicist Hermann von Helmholtz (1821–1894). Boutroux was a philosopher of science who took an anti-materialist position. Bergson's philosophy also bears several similarities to the ideas of the American pragmatist philosopher and psychologist William James (1842–1910). Bergson had sent him *Time and Free* Will and *Matter and Memory* but it wasn't until 1902 that James got around to reading these and sending Bergson a reply. James's radical empiricism was an attempt to reorient the empirical method by maintaining a proximity to direct experience. Similarly, Bergson's view on intuition focuses on the multiplicity of experience. At the same time he maintains that lived time should not be regarded as fragmentary: it is not a sequence of fixed moments. There is however also an important difference: whilst for Bergson, language formed an obstacle, since its structure determines the way we think about the world, for James immediate experience is already linguistic.

Intuition, from a scientific point of view, is not considered to be an exact method, since it precedes all epistemological conventions.[3] Stringent methods of observation, through quantification, measured experience, and self-critical, generalisable observation hold the key to produce knowledge that can be objectified. Any other approach to knowledge, in view of the rigorous methods of experimental science, are not sufficiently certain. However, intuition cannot be erased completely: the way one applies method is guided as much by intuition (e.g. the creative process of drafting a hypothesis) as by intellect. Bergson attributed a key role to intuition. As he explains in his *An Introduction to Metaphysics*, intuition could provide a way to experience the intrinsic being of an object's uniqueness and ineffability (Bergson 1912 [1903]). We not only relate to things via sense perception, but also intuitively. Bergson claimed for instance that we know our body "from without", via perceptions of both ourselves and others, but also "from within" (Bergson 1912 [1903], pp. 9–11), by the affections that we display, live and are aware of. It appears that neuroscience overly focus on the conclusions that can be drawn 'from without' (from a third person perspective), and do not sufficiently demonstrate awareness of the availability of an equally valid source of knowledge, namely knowledge from within. But, apart from direct experience, any form of consciousness is by necessity already memory (Bergson 1912 [1903]). After all, as soon as we become consciously aware of our experiences, experience itself has already occurred.

The role of intuition is important also in scientific ways of knowing, especially in the so-called context of (scientific) discovery. But this role is often polished away in *a posteriori* justifications of such knowledge. This retouch of science practices, reify a Whig history of a rationally derived process, as if reality itself speaks with a voice of reason, from the preconditioned, quarantined laboratory environment where it is given its podium: the elementary role of intuition is ignored as a-methodical and non-rationalist. But as Gilles Deleuze defends in his *Bergsonism*

---

[3] With the exception of the posterior acceptance of intuition as source of knowledge.

(1966), intuition is very much an exact method. It demands a concentrated self-critical assessment to remain open to intuition.

In his 'Specimen of the Table Talk' (1836) the poet Samuel Taylor Coleridge (1772–1834; also see footnote 6 of the previous chapter), inspired by a dictum of the German idealist philosopher Friedrich Schlegel (1772–1829), remarked: "Every man is born an Aristotelian, or a Platonist. I do not think it possible that any one born an Aristotelian can become a Platonist; and I am sure no born Platonist can ever change into an Aristotelian. They are the two classes of men, beside which it is next to impossible to conceive a third." (Colereridge 1836, p. 95). Slightly over a century later, the Argentinian writer and poet Jorge Luis Borges followed suit on this observation: "The latter [the Platonists] feel that ideas are realities: the former [the Aristotelians], that they are generalisations; for the latter, language is nothing but a system of arbitrary symbols: for the former, it is the map of the universe. The Platonist knows that the universe is somehow a cosmos, an order; that order, for the Aristotelian, can be an error or a fiction of our partial knowledge. Across the latitudes and the epochs, the two immortal antagonists change their name and language: one is Parmenides, Plato, Anselm, Leibniz, Kant, Francis Bradley; the other, Heraclitus, Aristotle, Roscelin, Locke, Hume, William James" (Borges 1945, p. 10). It is this latter tradition within which Bergson might also be categorised – specifically in view of his criticism of mechanistic and positivist thought, and his criticism of idealism.[4]

Parmenides adhered to the static conception of the world. His philosophy was influenced by the ideas of Pythagoras, and defended the view that the world is one substance, and this substance is eternal and unchangeable. In this view, our belief in change and plurality is deceptive and a product of sense perception. The world in itself is one. It is fixed. Heraclitus took a fully opposite position, claiming that everything is constantly changing, and opposites constantly coagulate and disperse again. He viewed the world as a series of converging and diverging processes. The profile of both modes of thinking can still be found in modern science, for instance: the static view in classical mechanics and the dynamical view in the theory of evolution. The view concerning two opposite positions that persist from presocratic Greek philosophy onwards might also be expressed in terms of their relation to time: either on the basis of the conception that everything is in stasis or on the basis of the opposite conception that everything is in flux. Of course, there are major differences between 'Plato versus Aristotle' and 'Parmenides versus Heraclitus', but some parallels may legitimately be drawn through history. Martin Heidegger referred to this duality when drawing a parallel between Bergson's position and

---

[4] The distinction can also be found in physics: the debate between Bohr and Einstein can ultimately be traced back to an opposition between these two traditions of thought. It even extends to antiquity in other thought traditions: Taoism and Confucianism, although both influencing the antique Chinese worldview, share a similar opposition, concerning fixedness or change, abstraction or concreteness, structure or process, principle or praxis.

Heraclitus, and Einstein's position and Parmenides's. Karl Popper perceived of a similar link, as did Bertrand Russell (Canales 2016, pp. 145 and further).[5]

Bergson sided with the dynamical mode of thinking, arguing like Heraclitus that everything is constantly changing, and in a constant state of becoming. His philosophy is a philosophy of change, a process philosophy. As such, it might be likened to certain aspects of Taoism in Chinese philosophy. In Tao, everything is also considered from the point of view of constant renewal. Taoism also considers the analytic mind, that attempts to discern reality by subdivision in its constituent parts, as limited. And Taoism also claims that the absence of thought is a precondition for freedom, rather perceiving of thought – identified with the first tense singular – as the throne of the free agent. Free will and determinism are equally useless points of view for a Taoist, since they presume some kind of impossible outer perspective, and the same counts for Bergson (as will be discussed at the end of Chap. 5). And thus, it should not be surprising that a central line in the philosophical writings of Bergson was his persistent criticism of mechanistic determinism – the worldview that emerged during the scientific revolution in the seventeenth and eighteenth centuries, resulting in the watchmaker's view cited above this chapter. This watchmaker's view of the world was the dominant position in classical science (of which Isaac Newton's (1642–1727) theories are the most well-known example), but became increasingly complemented and perhaps even eclipsed, also in science, with a view that embraces process and organic evolution.

The philosophy of science aims to define proper ways of reasoning and correct forms of argumentation. The field focuses on the definition of norms for verification – and later falsification – of hypotheses, as well as norms for the proper ways to conduct empirical research, draw argued conclusions and thus ways to further the expansion of knowledge. Scientific practice itself however often follows different paths. Experiments do not merely fulfil the need for methodological confirmation of hypotheses. They are in the first place demonstrations in the sense of are like theatrical performances (Toonders et al. 2016). This role is necessary, since novelty is often discouraged by the social contexts of established expertise: renewal in specific knowledge fields often proves to occur only with the pension age of a mentor generation rather than with the success of a creative new approach to the subject matter at hand. Still, 'premature' paradigm changes do occur, and they could be furthered by investigating the presuppositions that are implicit to specific fields of research and their established theoretical expertise. This in itself is an important legitimation for the existence of the philosophy of science. In this vein, the focus of this book lies on explaining the relevance of Bergson's ideas not in the first place for philosophy, but especially for science. To this aim, it gives an account of the dialogue between

---

[5] The divide might also be associated with the German idealist tradition in comparison with the British empiricist tradition: the abstract versus the concrete; the realist interpretation of symbols and concepts versus the nominalist conception of these; the intuitive versus the rational; the poetically Gaelic versus the square Teutonic. But this of course amounts to an overly naive Romantic reductionism of thought and culture to nationality and language.

Bergson and the scientists of his age and applies the insights gathered to diagnose contemporary discourse in science and technology.

Although the oeuvre of Henri Bergson addresses a broad range of interesting topics,[6] I will focus on three core motifs: *duration (durée)*, *vital impulse*[7] *(élan vital)* and *immediate concreteness*.[8] Gilles Deleuze (1991 [1966]) focused on a slightly different tripartite division: the motifs of duration, vital impulse and memory. Here I opt for an alternative approach because my three motifs are directly related to the three scientific domains that will be addressed in this book – physics, the life sciences and the neurosciences. Although they are widely differing fields of expertise they share a common denominator in their decisive impact on our self-understanding and our understanding of our place in the world. Furthermore, they remain at the core of contemporary debates on the impact of science on society: they present themselves with an opportunity to see how science might contribute to a diagnosis of the human condition. It is for this reason that this book is written.

Immediate concreteness lends itself best for discussing neuropsychology and neuroscience – memory would narrow down the scope of the debate – while *vital impulse* allows me to discuss the life sciences and *duration* to discuss physics. These links are not exclusive, however: the concept of duration for instance not only applies to physics, but to life and memory as well. And the concept of vital impulse might be applicable to epistemological complexity in science: scientific knowledge systems, after all, are also often understood in terms of an evolution of systemic knowledge spheres. Similarly, the issues related to immediate experience, are relevant for both epistemological and neuroscientific explanations of perception. This book focuses on three knowledge areas that are treated in both their history and their current state of the art. It diverges from the tripartite division suggested by Giles Deleuze, who, in his reading of Bergson, suggested a focus on time, on life and on the virtual.

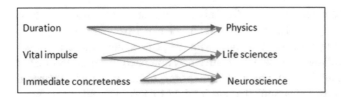

There is another reason to take the three concepts as leading concepts for a book. They offer a discursive alternative for a philosophy of science that continues to be haunted by René Descartes's dualistic worldview. They create a vocabulary that

---

[6] Some possibilities might include Bergson's discussion of creativity or imagination

[7] Sometimes also translated as 'vital impetus'.

[8] In the next chapter I will discuss these in reversed order since this better enables me to discuss them in relation to Bergson's thought. In the priotorisation of the chapters after Chap. 2 however, their prioritisation remains in the order in which I mention them here, although, as shown in the figure here, they are interrelated.

supersedes binary opposites between 'me' and 'it', 'we' and 'all other organisms', 'here' and 'the rest of the universe' – a vocabulary that inevitably leads back to dualist Cartesian metaphysics. This dualism is, according to Bergson, in his introduction to *Matter and Memory*, "though suggested by the immediate verdict of consciousness and adopted by common sense, to be held, in small honour among philosophers" (Bergson 1911 [1907], p. xii). The dualist position is, however, in many cases unavoidable, specifically where knowledge systems need to take it as a precondition of knowledge as such, but its limitations should remain articulate enough to remain alert to possible bias.

Vital impulse is duration (Bor 1990). They are two sides of the same 'thing'. And they might even be aligned to immediate experience. And indeed, it is tempting to conflate the three concepts of *immediate concreteness*, *vital impulse* and *duration* and subsume them under the heading of 'fluidity': the ever-changing ever-evolving dynamics. This conflation would then cover how 'things'[9] in reality constantly reveal themselves in processes of convergence and individuation. But I am reluctant to do so: first of all because these three interrelated concepts allow me to critically assess the three scientific domains mentioned above (neuropsychology, life sciences and physics). Nonetheless, in the course of the analysis I will point out that these three concepts are indeed interrelated, as basic ingredients of Bergson's process philosophy, for all three concepts convey the idea of continuous change, although they unwind this notion of change in diverging directions.

As mentioned, Bergson's notion of immediate experience is, as radical empiricism for William James, is at the core of his philosophy (Bor 1990). This is an exceptional view for a philosophy of science, because most philosophers of science agree that, in general, immediate experience, preceding even introspection, is not a reliable starting point for scientific research. Rather, science aims to produce and analyse phenomena in the controlled artificial setting of the laboratory, which demands not a focus on uniqueness but on patterns and similarities that can be generalised. As Bergson wrote in *Creative Evolution* "in dealing with things science is concerned only with the aspect of repetition" (Bergson 1911 [1907], p. 45). Conducting an experiment basically means: repeating a particular event as often as possible under various controlled condition, on the basis of a hypothesis; comparing the results; recognising similarities; generalising these in temporarily valid laws. Although not all unique phenomena are excluded from the scientific gaze (for instance, the origin of the universe) such single events are broken down by scientists

---

[9] Put in inverted commas here, since 'things' appear to refer to a definition of objects (stones, trunks, utensils without known functions), separate and outside of time and processes. But although the word 'thing' might indeed have such connotations, the word originally referred to something opposite to these connotations. It meant "what is under consideration", meant 'meeting, assembly, council, discussion' (e.g. as it persists in the name for the Icelandic parliament: the Alþingi, or Althing. Only later it came to mean "entity, being, matter". The original word stem might have referred to 'appointed time', since the root –"t(h)in - is taken to mean 'stretch', and is supposed to have gradually come to refer to a stretch of time, for a meeting or assembly". Bergson's philosophy thus seems to follow the opposite path from the etymological evolution of the word thing (he went from substance to duration rather than from duration to substance).

into various aspects which are partially repeatable, allow for comparison, scientific analysis etc.. In this regard, unique events cannot be studied from a scientific point of view. Thus "[a]nything that is irreducible and irreversible in the successive moments of a history eludes science" (Bergson 1911 [1907], p. 45). Essentially, each event is unique, but not all its features are.

Although often explicitly anti-metaphysical in its pragmatic stance, modern science does entail an implicit metaphysics. For science, events necessitate framing in a wider context of comparable instances. It is only by observing clusters of phenomena – repeatable and comparable – that predictions become possible. The interpretations of scientific observations are therefore based on repeatable phenomena, mediated by assumptions and conventions. There is an implicit metaphysics at work here, for according to the scientific method, natural events are repetitive rather than unique, a claim that precedes empirical experience, rather than following from it. Such hidden metaphysical claims entail important consequences for the research fields in question and the insights they produce.

Ontological claims predetermine its interpretations and conclusions. Some of these form the very backbone of science: the principle of causality; the separation between subject and object; the conviction that organic and inorganic entities[10] are fundamentally speaking of the same nature etc. Some further examples include the notion of time as fourth geometrical dimension, the view that living organisms function like machines, or the brain-centred view of the mind ('we *are* our brain'). These and other metaphysical assumptions involved often remain concealed, so that, as Bergson states, they are mistaken for scientific conclusions.

The question from which this study departs is: *how can the Bergsonian revision of the notions of time, life and memory provide for an ontological basis for a philosophical approach to the technological nature of the human being in its current habitat?* This question necessitates a study of the different implicit ontologies in the three science domains sketched out, as well as the outlines of an analysis of the current technological condition on its basis. According to Bergson, science should not try to purge itself from metaphysics and to rid itself from ontology. Metaphysical perspectives are always and inevitably included in scientific research practices. Scientific research always emerges against a backdrop of a priori metaphysical conventions, assumptions and convictions, even if we do not always recognize them as such. Implicit metaphysical assumptions cannot be removed, and thus, it would be more beneficial to the quality of scientific knowledge to articulate such assumptions in order to better discuss and question their nature, function and impact. After all, progress in scientific research cannot be outsourced to an automatic process. Interpretation on the basis of preconception will always be needed to make sense of the data generated. But if unchecked and unrecognised, such conditioning risks being presented as an objective conclusion that was generated by observations, rather than forming an a priori frame of interpretation for these observations. In

---

[10] Here defined as living and non-living, not as in (bio)chemistry where there are of course also nonliving organic molecules and structures.

other words: a priori is too often mistaken for a posteriori. The metaphysical positions that shape the conventions of science, and generate the concepts used in science need to be adapted to experience (Blanco-Pérez 2018). They should not shape experience. Metaphysics should therefore not be left merely to the philosopher: critical articulation of hidden assumptions should always also be a scientists responsibility. But before running ahead of things, we will need to determine what we mean by the term 'metaphysics', and what type of metaphysics Bergson adheres to.

Literature

Bergson, H. (1912 [1903]). *An introduction to metaphysics* (trans; Hulme, T.E.). Hackett Publishing Company 1999. Also included in: Bergson, H. (2007 (1923)). *The creative mind: An introduction to metaphysics* 1923. Dover Publications: Dover.

Bergson, H. (1911 [1907]). *Creative evolution*. New York: Henry Holt and Company.

Blanco-Pérez, C. (2018, Julio-Diciembre). Henri Bergson: A solution to the mind-body problem. *Miscelanea Comillas, 76*(149).

Bor, J. (1990). *Bergson en de onmiddellijke ervaring*. Meppel: Boom.

Borges, J. L. L. (1945). *Nota preliminar, en W. James Pragmatismo. Un nombre nuevo para algunos viejos modos de pensar*. Buenos Aires: Emecé.

Canales, J. (2016). Einstein's Bergson problem: Communication, consensus and good science. *Boston Studies in the Philosophy and History of Science, 285*, 58.

Colerigde, S. T. (1836). *Specimen of the table talk* (2nd ed.). London: John Murray.

Deleuze, G. (1991 [1966]). *Bergsonism* (trans. Tomlinson, H., Habberjam, B.). New York: Zone Books.

Reichenbach, H. (1984 [1956]). *The direction of time*. Mineola: Dover Publications.

Toonders, W., Verhoeff, R., & Zwart, H. (2016). Performing the future: On the use of drama in philosophy courses for science students. *Science & Education*. https://doi.org/10.1007/s11191-016-9853-3.

Chapter 2
An Attempt to an Applied Metaphysics

Bergson's position in metaphysics might be clarified by relating it to the problem of the universals, a theoretical struggle in the late middle ages. Most philosophers revisit the problem of the universals only reluctantly. Also Bergson only refers to the problem of the universals once (in *Matter and Memory* (1988 [1896], pp. 202–212). Still, it is – as Bergson biographer Milič Čapek's elaborate discussion of the issue (1978) illustrates – important for understanding his work and his critical stance towards the symbolical (language, semiotics etc.). It doesn't merely provide us with the ability to place his oeuvre in the wider context of the history of philosophy. It also clarifies how it relates to the 'linguistic turn', which has dominated philosophy from the 1950s to the present.[1] Further, the debate on universals concerns the reality status of concepts versus the reality status of real things. This is relevant since at times, theoretical science is not sufficiently aware of the status of its own conceptual contents, as Bergson argues. And this was to become one of the driving forces of the controversies between him and Einstein.

In the debate on the universals, conceptual realists, who believed concepts had a reality outside of human existence, were pitted against nominalists, who adhered to the position that concepts only exist in name, in language: realists, building onto the works of Plato, took the position that reality has an inherent structure which can be discovered and described. Although some descriptions may not yet be correct nor complete, eventually we will be able to uncover such structures, and as such 'carve nature at its joints'.[2] Nominalists, building onto the works of Aristotle, believed that universal forms are merely concepts: they do not have a real or autonomous status. Reality is too pluriform, and indivisible. It does not present itself as representable. And names are merely names, they are not realities in themselves.

Stat rosa pristina nomine, nomina nuda tenemus. This sentence translates as 'the rose of old remains only in its name; we merely possess naked names', or more literally: 'the pure [pristina] rose [rosa] stands/persists [stat] in name [nomine], we have/are left with [tenemus] the naked [nuda] names [nomina]'. This adage was composed by the Benedictine priest Bernard of Cluny, in his *De Contemptu Mundi*

---

[1] The linguistic turn was spurred by the idea that our experience of the world is always mediated by the language we use, and thus our perception of the world is always determined by language. Such determination however, combined with the arbitrary shape signs have taken, also implies that we cannot look past the horizon of language, since our ways of thinking are restricted to its arbitrary shapes. Before the turn, language could still be regarded as an instrument of thought. This view exists throughout the philosophical world; it is also voiced by Confucius, who stated the desire to rectify names to make them correspond to reality (Confucius, *Analects*, Book XIII, Chapter 3, verses 4–7, translated by James Legge (1893)).

[2] Plato: *Phaedrus* 265d-266a. For an English translation see Nehamas and Woodruff (1995).

© Springer Nature Switzerland AG 2021
L. Landeweerd, *Time, Life & Memory*, Library of Ethics and Applied Philosophy 38, https://doi.org/10.1007/978-3-030-56853-5_2

(Cluny 1906 [twelfth century]). In this work he criticised a church in Rome that has sunk to become Rome in name only. It is also the last sentence of Umberto Eco's well-known novel *The Name of the Rose* (1980). The rose in Bernard of Cluny's adage was a core metaphor in the problem of the universals. Dealing with the problem of the relation between universal forms and particular things, its main issue was that although we might have concepts, they do not necessarily refer to reality in its pluralistic nature. Bernard's adage is one of the most elegant summaries of the 'problem of words and things', which in metaphysics has never been truly solved. Looking at the nature of things, philosophers assumed two positions for generations: either it was assumed that forms, in the Platonic sense, were decisive for the emergence of individual things, or it was assumed, from an Aristotelian perspective, that our knowledge is based on individual specific properties or relations (Brougham 1993).

The struggle over concepts was a clash between realists, prioritising concepts and forms as attributable to reality, and nominalists, prioritising the unknowable existence of individual things, and regarding concepts as arbitrary and relative to us. Aristotle rejected the idea that names are derived from things: "'Man', and indeed every general predicate, signifies not an individual, but some quality, or quantity or relation, or something of that sort". (Aristotle 1984). Nominalists agreed with Aristotle that Plato's forms do not have an independent existence. Thus names are nothing but our categorisation of individual things in their actualised existence. Nominalism is another term for conceptual formalism: concepts are merely a form of understanding the world, they do not relate to something real in nature. They are part of our way of referring to the world, they do not require being part of that world.

The example of the rose is often taken to explain the issue: what's in a name? A rose will smell as sweet, regardless of our labels. The most famous reference to the rose, a covert pun on the dilemma, can be found in Romeo and Juliet. To cite Shakespeare's Juliet in an allusion to the debate, the question is: "Would a rose smell as sweet if called by another name". For the conceptual realist, the rose, the real thing, can be captured by its form, its name, at least, if correctly phrased. For the nominalist, any individual rose would smell sweet apart from the names we attribute to it. In the words of Romeo's later response: "What's in a name? That which we call a rose, by any other word would smell as sweet." Romeo and Juliet are obviously well versed in nominalism. But for the realist, it is the form that decides whether the rose smells as sweet as it does. So, if it had been up to those of a Platonic, conceptual realist, persuasion, Romeo and Juliet would have had to remain enemies, since Montegues and Capulets remain defined by their universal form – their belonging to the categories of their respective families.

But, speaking from a realist point of view, these labels cannot be dismissed with as merely hollow instruments of sophistry. Our representations are in themselves part of this indivisible world. They are directly connected to the material world, in terms of their relation to instruments, tools, discursive practices, and artefacts. The medieval debate on universals persisted, albeit in a somewhat different guise, in the modern distinction between rationalism and empiricism. These two schools of thought would converge in the modern scientific method.

The debate of the universals often seems to disappear from the agenda, but it has the tendency to pop up unexpectedly and in different guises. One such guise is the linguistic turn in philosophy, that occurred from the 1930s to the 1970s. This turn implied an ever-growing focus on the structure of language, and the way in which this structure is all-decisive to notions of validity. In its later phases, language came to be considered fully self-referential. In this regard, this later phase, often loosely referred to as 'postmodernism'[3] can actually can be regarded as a resurgence of nominalism. Conceptual realism, or conceptualism, in its medieval meaning, did influence some proponents of the rationalist tradition in philosophy in prioritising concepts, forms and self-derivable arguments over concrete things. As such, rationalism stood against empiricism in prioritising the sensory experience of concrete things over concepts and a priori reason. And in a sense, this struggle persisted, since science still sometimes is guilty of, paraphrasing William James, 'the trick of turning names into things' (James 1975 [1907). But there is a trick involved in the study of names as things as well. As nominalism became overly involved in the structures of concepts, so too, the linguistic turn that came to dominate post WW II western philosophy entangled itself in the structures of language. As the early twentieth century Indian poet and writer Rabindranath Tagore phrased it in his book *Sadhana*: "The men who are cursed with the gift of the literal mind are the unfortunate ones who are always busy with the nets and neglect the fishing" (Tagore 1915, p. 72). And as such, the postmodern fisher king remains overly preoccupied with the idea that concepts are real, and the real is merely concept. In that case, philosophy should be about words only, not about things.

Bergson's philosophy is not a form of nominalism, but it did not adhere to the principles of realism either. Bergson's view was that nominalism, prioritising individual things, cannot avoid being reduced to conceptualism. It cannot avoid creating classes, or genera, and presupposing that these are realist in nature. At the same time, conceptualism, prioritising classes over individual things, and taking these classes to be real, cannot avoid being reduced to nominalism, since it needs presuppose individual things. As Bergson stated: "The first [conceptualism] composes the genus by an enumeration; the second [nominalism] disengages it by an analysis" (Bergson 1988 (1896), p. 205). And as such both conceptualism (the realists) and nominalism rely on what Čapek refers to as 'atomistic sensualism' – the position taken by David Hume in which thought is reduced to a kaleidoscopic mosaic of 'sensory elements' (Čapek 1971).

The problem with atomistic sensualism is that it fails to acknowledge the living continuity that exists in our inner experience (Buford and Oliver 2002). The opposition between nominalism and realism inevitably leads to such a view. The point Bergson tried to make was that thought is a dynamic process rather than a static series of mental images. It exists somewhere within the spectrum that lies between that which would be only word, or that would only be an individual sensory

---

[3] Whilst the term 'postmodernism' originated in architecture, to designate an ironic and pluralistic approach to style, it came to designate a series of movements, schools and ideas ranging from art to philosophy, from literature to sociology.

impression. Bergson opted for neither a prioritisation of individual things over forms, nor a prioritisation of forms over individual things. He opted out of the acknowledgement of an atomism of things as well as an atomism of mental impressions. His 'vitalism' allows for a position that is neither nominalist nor realist. But as such, his work reopens a debate which the modern scientific method allegedly solved.

Bergson himself did not write much about the battle of the universals. Still, Čapek took his frugal remarks on the matter as the starting point of his discussion of Bergson's philosophy, notably because of Bergson's intuitionist conviction that we can never 'know' the individual and unique nature of a thing and can only intuit it. The problem of the universals is relevant here for several reasons: the exact sciences often use models to study certain aspects of the phenomena they are interested in. But in some cases, the status of these models is forgotten: thus, atom models became realities, outlining the structure of the smallest unit of the universe, clock time became real time, and neural networks became the mind. But models should be regarded as conceptual instruments, as tools rather than as truths. Here, the battle of the universals persists, in a lack of acknowledgment of the distinction between the real and the symbolic. Its relevance is that we should be cautioned against mistaking the abstract and the conceptual for the concrete.[4]

For Bergson, science and metaphysics designate separate (although interacting) activities of the mind. Bergson insists on a difference between conceptual thought and intuition. Science makes use of conceptual thought. It is the instrument through which it generates knowledge about matter. Metaphysics makes use of intuition, to be able to know 'spirit' (Bergson 1912 [1903]), Sfara 2015). There are thinkers who discuss knowledge in terms of objectivity and rationality (Leibniz, Comte, Poincaré) and philosophers who discuss knowledge in terms of subjectivity, intuition, and sensory experience (Nietzsche, Sartre, Merleau-Ponty). Kantian philosophy attempted to wed the two, and whilst Bergson might be fitted within the second category, his aim was to show how subjective experience and intuition formed a necessary complement, rather than an alternative, to the rational intellect. And reasoning from both, Bergson's opposition between matter and spirit might remind us of Platonic distinction between the real and the ideal, or the Cartesian distinction between the substance of matter and the substance of mind. But we should avoid such an interpretation of Bergson's metaphysical views. Spirit holds a different meaning in his oeuvre than either the ideal or thought. He describes the distinction between spirit and matter as temporal in nature, rather than in conceptual (real/ideal) or substantial (two substances) terms. To put it in Bergson's terminology: I have an intuitive access to the temporal nature of direct, non-mediated experience. This access connects me to a continuity of diverse and multiple durations. If I put in the right effort, I can follow this direct experience either to spirit or to matter (Bergson 2007 [1923] p. 187). Here, Bergson comes eerily close to a Cartesian

---

[4] Another process philosopher, Alfred North Whitehead, coined this figure of speech in reference to his definition of the fallacy of misplaced concreteness.

point of view, and indeed, he ultimately does consider himself a dualist – but, as we will see – of a different kind than the one arising from a superficial interpretation of Descartes ideas.

In this chapter, I will describe more in depth the three interrelated motifs that I drew from Bergson's oeuvre to inform the further discussions in this study: immediate concreteness, vital impulse and duration. They all relate to a dynamic and fluid worldview, as opposed to a static and analytic one. The concepts are interwoven and reoccur throughout the writings of Bergson. The three motifs all treat the following question: how is 'discrete individuality' possible whilst direct experience is always in flux, indiscrete, and holistic? The main point of this reflective and reflexive exercise is to use them as notions for a concretisation of Bergson's philosophy for contemporary science, first through science and subsequently through a discussion of technology, in relation to the self-creative, or autopoietic, nature of human evolution and its impact on global ecosystems.

2.1   Three Notions for an Applied Metaphysics

Bergson's oeuvre consists of writings on a broad spectrum of issues. He addressed the notions of evolution, time, instinct, intuition and reason, he developed a theory on the origins of morality and religion, a theory on aesthetics, a metaphysics, a theory on laughter and wrote a tractate on free will. He provided for a critical analysis of psychology, a criticism of consumerism and a he wrote a work on mysticism. Bergson defined metaphysics as the attempt to, in Ansell-Pearson's words, think beyond the "acquired and sedimented habits of the human mind" (Ansell-Pearson 2018, p. 75). According to Bergson, these habitual ways of thinking tend to be influenced by mechanistic and spatial conceptions of the world. Essentially, the view on the world as spatial and geometrical is metaphysical in nature, but it is underarticulated in its influence, steering certain cognitive automatisms, in science and in daily life alike.

About the enormity of Spinoza's work, Bergson once wrote: "One might say that all philosophers have two philosophies, their own and that of Spinoza"[5] (Bergson 1927a, b). Still, Bergson was critical of Spinoza's geometrical ordering of the world. He rejected mechanistic determinism, whilst ultimately, Spinoza's philosophy remains determinist. To illustrate Bergson's opposition to mechanistic thought, one might refer to the principle involved in the assumption of the (ir)reversibility of the arrow of time. Within physics, it is assumed that although time itself is like a geometrical category (either taken as linear and absolute or as relative), processes that take place in time might be reversed in such a way that, with a randomly chosen beginning-state and end-state, if one reverses the forces responsible for going from

---

[5] In an often quoted letter to Léon Brunschvicg: "Tout philosophe a deux philosophies: la sienne et celle de Spinoza."

beginning to end, one will have a state that is identical to the initial state. Often, the metaphor of a game of snooker, or, better even, billiards is taken to illustrate this point: the cue hits a specific point on the ball with a certain force and direction; the ball rolls until it stops. The cue again hits ball, with the same force, but now hitting the opposite side of the ball, and lo and behold: the ball will wind up at its original position. Now: disturbing elements of this process come from outside. The felt of the table or the top of the cue may be slightly irregular, the temperature may vary etc. But as far as classic mechanics is concerned, these variances are merely disturbances of what ultimately remain universally valid and inescapable laws that govern the universe.

Bergson saw reality as something that constantly renews itself. He refused to take the Spinozist position that these 'renewals' are predetermined by earlier states. In contrast with Spinoza's account, he did not embrace the idea that these changes were manifest through a mechanism that equated future states with past states. He succeeded in offering an alternative that at the same time safeguards our notion of human freedom by a rethinking of time. This necessitated a prioritisation of process over substance and possibility over actuality.

As broad and eclectic as Bergson's publications may seem, they all bear a relation to the three central philosophical notions mentioned. They frame this book's interpretation and application of his philosophy: (a) the possibility of immediate experience of how the world is concretely given (French: 'données immédiates'), (b) the concept of vital impulse, or impetus (French: 'élan vital'), and (c) duration (French: 'durée'), as distinct from clock time. On the basis of these three notions, Bergson developed a process metaphysics. All three are strongly intertwined and all three relate to the three science domains discussed in the ensuing chapters: physics, the life sciences and neuroscience. I will discuss these in the ensuing chapters, ultimately narrowing down on the nature of technology and on the role of these domains in currently emerging technologies. For those who are less inclined to theoretical philosophy but still picked up this study: some basic understanding of these three concepts is necessary to follow the vocabulary and the reasoning in the ensuing chapters.

2.1.1 Immediate Concreteness

Immediate concreteness, or Immediacy is a notion central to Bergson's doctoral thesis *Time and Free Will* (1889). The French title *Essai sur les données immédiates de la conscience* (added as subtitle in English as: 'essay on the immediate data [literally: 'givennesses'] of consciousness') shows what underlies these notions of time and free will: direct experience, without mediation by concepts, conventions, or even the way in which the nature of our sensory apparatus frames experience. The latter might sound strange, but their nature is something that we can only assess a posteriori, it is not given in the direct data of consciousness. The notion of immediacy shows a reluctance to accept one of Immanuel Kant's basic premises: that the world of things as it exists upon itself (das Ding an sich) is beyond our experiential horizon. Bergson was 30 at the time, and his thesis already set the stage for the other works that he was to publish.

In *Time and Free Will*, Bergson sketches out how our knowledge of what is given in our experience of our consciousness is fundamentally different from our knowledge of 'objects' in the world. We should not assume that our inner states of awareness can be studied in the same way as we study phenomena external to our consciousness. We cannot measure these states, we cannot subject them to controlled laboratory experiments without construing them as external objects, thus blinding ourselves to their nature. They should be understood 'in their developing, and in so far as they make up, by their interpenetration, the continuous evolution of a free person (paraphrased from Barnard (2012) in his discussion of pp. 229 from *Time and Free Will* (2001 [1889]). They reveal themselves in their concrete nature. As a consequence, the age-old dilemma between determinism and free will is merely the result of a confusion caused by an illegitimate theoretical relation drawn from the 'un-extended' into the 'extended' (Bergson 2001 [1889] pp. xiii and 70). The dilemma is unfounded, since it does not articulate the conceptual status of these two basic notions. But as said, access to immediately given data is not something that is self-evident. Although we constantly and experience, in the immediate sense, we are not constantly consciously aware of this.

Bergson refers to 'the good sense' (*le bon sense*), as an attitude that demands a sacrifice of our convictions, as these may lead to intellectual automatism, mere common habituation. To enable this, one has to labour to rid one's mind of one's habits of thought, one's predispositions and the conventions that rule one's way of being aware. In other words, one has to go beyond the mere anthropocentric position of humanism. Furthermore, one should not mistake this type of consciousness with its various conceptual representations. And essentially, only one's experience of self is immediate.

Bergson's most important contribution to the philosophy of science might lie in his observation that the immediate ways in which we are aware of the data of our consciousness are deeply intertwined with time. Bergson's distinction between two concepts of time (*temps* and *durée*) stems from a specific observation he once made that time as we experience it has a duration whilst time as we measure it does not: mechanical, measured time thus conceptualises time as something fundamentally alien to time as something concrete, experienced, something that is undergone. And it is only in the latter that we experience immediacy. The former merely construes it in a confusion between measurement and what is measured. But a clock does not define time, nor does an astrolabe, in the measurement of our position of the stars, determine the existence of spatial position.

In our experience of the world, sense data continuously bombard our consciousness, and we tend to make sense of such data predominantly on the basis of visual templates. This inevitably prompts us to interpret the world in terms of distinct objects, far away or nearby, large or small. Whilst our sense of touch underscores this sense of solidity of the material things that surround us, hearing provides us with a very different notion of the world: here we discern a world that is deeply intertwined, and continuously evolving. A well-known passage (a.o. quoted by Barnard 2012) from Bergson's alternative intuition of the world can be found in his *The Creative Mind; an introduction to metaphysics* (2007 [1923]):

> "Let us listen to a melody, allowing ourselves to be lulled by it: do we not have a clear perception of a movement which is not attached to a mobile (i.e. a moving) object, of change without anything changing? This change is enough, it is the thing itself. And even if it takes time, it is still indivisible; if the melody stopped sooner, it would no longer be the same sonorous whole, it would be another, equally indivisible." (Bergson 2007 (1923), p. 174).

Our immediate access to the given data to consciousness is not solely about consciousness, but very much about the world. It does not need the presupposition of any 'thing'.

All individual things are singular and as such they escape definition. Defining, here, should be understood in the Aristotelian sense: subsuming specific specimen under more generic classes. But single events and single entities remain indeterminable. *Individuum est ineffabile* – the individual cannot be known. This ancient Latin, and originally Greek, adage refers to the impossibility to give any definitions or make any claims about concrete things. Any knowledge is always knowledge of some generalised class of things. It can only apply to similarities between different things. We tend to think that reality is always individual, and that reality thus consists merely of a kaleidoscope of substances. What it is, never presents itself in terms of a whole. Thus, we tend to neglect the continuous nature of things, the way in which the world dynamically unfolds.

In 1911 Bertrand Russell attended a lecture by Bergson at University College London on the 28th of October, and on the 30th of October Bergson attended one of Russell's lectures Bergson (Vrahimis 2011). Russell regarded Bergson's convictions concerning the nature of numbers as naive (Russel 1912; Petrov 2013; Čapek 1971). Russell read Bergson's approach as and approach that confuses numbers with their concrete aggregates. But is the very notion of referents and aggregates that is dismissed with ex ante in Bergson's work. Bergson's aim is not to criticise mathematics, logic or empiricism, but the way in which these are interpreted in mainstream discourse. His approach to empiricism both criticises numerical views on time and holistic views on experience. Whenever the multiple nature of time is revealed in our experience as duration, it appears as indivisible: it is not analysable in its different components. Russell neglected Bergson's stated difference between numbering and counting, and as such criticism appears to consist of a selective reading of Bergson's texts (Petrov 2013; Čapek 1971).

Counting, numbers and multiplicity are tied up with our notion of space and spatiality. Space is homogeneous, quantitative, and actual. Duration however is heterogeneous, qualitative and virtual[6]:

---

[6]The distinction between actual and virtual will be discussed in Chap. 5 but suffice to say that the virtual is related to memory, imagination, projection, possibility etc.

"When the mathematician calculates the future state of a system at the end of a time t, there is nothing to prevent him from supposing that the universe vanishes from this moment till that, and suddenly reappears. [...] In short, the world the mathematician deals with is a world that dies and is reborn at every instant – the world which Descartes was thinking of when he spoke of continued creation." (*Creative Evolution* 1911 [1907], p. 39).

But to Bergson's mind, this view on succession remains too much attached to the spatial: it leads to a thinking of duration as a mere succession of instances.

Describing duration in terms of the act of counting, a succession of fixed moments, like frames in celluloid, would not be adequate, since processes should be regarded as a continuum. But how can one safeguard multiplicity in our conceptualisation of time whilst on the one hand avoiding holism, and on the other hand avoiding the notion of a succession of fixed instances? In 'Time and Free Will', Bergson provides a definition of the nature of numbers: "Number may be defined in general as a collection of units, or, speaking more exactly, as the synthesis of the one and the many" (p. 75). And further: "[...] the idea of number implies the simple intuition of a multiplicity of parts or units, which are absolutely alike. And yet they must be somehow distinct from one another, since otherwise they would merge into a single unit" (pp. 76–77). Bergson poses that the answer lies in the awareness of the concrete, to which numbers refer. As Gilles Deleuze states, for Bergson: "duration divides up and does so constantly: that is why it is a multiplicity. But it does not divide up without changing in kind...: that is why it is a nonnumerical multiplicity, where we can speak of 'indivisibles' at each stage of the division" (Deleuze 1966 (1991) p. 42).

To illustrate Bergson's quest for a balance between multiplicity and continuity, the above study of movement and light was painted in 1921–1922, by the Czech futurist artist František Kupka (1871–1957). Kupka's works and writings were influenced by Bergson's *Matter and Memory* and *Creative Evolution* (Leighten 2013, p. 164).[7] The above study depicts a fun fair; or rather, Kupka's analytic representation of his perception of movement at a Parisian fair (Image 2.1). Kupka was a pioneer of abstract painting, although his blurred kaleidoscopic works do hide an indirect realism: if one looks long enough, one can make out the shapes of people, their movement fragmented in a series of changed shapes. Kupka's painting analyses movement, dissects it into a series of fragments of perception: the paradox illustrated is what moved Bergson to articulate his view that duration is not properly articulated in our habitual representations of time: whenever change is thought, or represented, it is done so through its opposite: a freezing of time. To Bergson, this problem was determined by the language we use, and the way it lures us into a confusion between concept and reality.

As I discussed earlier in this chapter with regard to nominalism and realism, the separation between abstract and concrete (or concept and reality) is a central line in Bergson's thought. In his view, the process nature of things can be experienced

---

[7] As might be illustrated by the following quotation from Frantisek Kupka's *Creation in the Plastic Arts* (1989): "Art expresses itself in composing its own organism. The work of art possesses a specific organic structure, entirely different from that which is found in nature."

Image 2.1 František Kupka *La foire ou La contredanse* (The fair or The contredanse) Oil on canvas 73 × 328 cm 1921–1922

directly, but the things themselves can only be known indirectly, via concepts, symbols, comparisons, repetition, etc.. 'Awareness', to put it in a psychological term, exists somewhere between the immediate data of consciousness and the mediated nature of thought. These two poles were already introduced in Bergson's doctoral dissertation *Time and Free Will* (1889), as the distinction between 'la conscience immédiate' and 'la conscience réfléchie'. This subject would remain central throughout Bergson's career (Bor 1990). Immediate consciousness is given in sensory experience, whilst our reflection on that sensory experience is organised in such a way that it normally organises the world into discrete sets of two, a dichotomous organisation of yes or no, absent or present; 1 or 0, biotic or abiotic etc.

On superficial inspection, Bergson's reflected consciousness seems to overlap with Kant's notion of apperception (represented as an accompaniment of the 'I think'), whilst Kant's notion of (internal) representation seems similar to Bergson's immediate consciousness. There is however a very important distinction between Bergson and Kant regarding the latter. Whereas Kant defines representations as, in a way, the conscious aspect of sensory perception, for Bergson, sensory perception in itself is already conscious (albeit in terms of memory). His critical departure from Kant thus concerns the transcendental aspect of his epistemology. However, whilst on first sight, Bergson's writings appear as starkly anti-Kantian, his project can also be read as an attempt to reintroduce the Cartesian notion of intuition into the epistemology we were left with after Kant (Riquier 2016), an epistemology that regards space and time as empty forms through which phenomena are apperceived.

Bergson, in his critical revision of Kantian philosophy distinguishes between immediate experience on the one hand and reflected consciousness on the other. Here he follows another basic structure of Immanuel Kant's epistemology. Kant defined several forms of rational judgement. The point at which it is tempting to draw a parallel between Kant and Bergson lies in the knowing subject's ability to perceive his own perceptions: this is the transcendental argument of apperception. Apperception can be regarded as the apprehension of a (mental) representation, which in its turn derives from perception. Apperceptions are always automatically seen as 'one's own'. This does not exclude the possibility that they can be the basis of general judgements, but epistemologically speaking, they are 'of the subject'. In

Kant's words (§16 of the *Kritik der Reinen Vernunft*): "It must be possible for the 'I think' to accompany all my representations; for otherwise something would be represented in me which could not be thought at all, and that is equivalent to saying that the representation would be impossible, or at least would be nothing to me" (Kant 1929 [1781/1787(B)] pp. 131–132)".

For Kant, space and time are not aspects of reality, but *our* way of experiencing reality. Succession in time is a category of experience, not of reality. But time for Bergson is not a linear process of succession, either experienced or real. Time is concretely and immediately given, and not as a succession of temporal points, but as flux, as stream of change. This alternative view on time makes the Kantian 'Copernican turn', in which the knowing subject constitutes the world as he knows it, obsolete.

For Kant, we experience the world of nature, of physics and causal relationships, as a temporal sequence in the linear sense of the word. However, because this does not apply to the world in itself but rather our experience of it, there is still room for free will and autonomous decision-making Kant's worldview. The Kantian reconstruction of free will also came at a price: the need to consider all knowledge as relative to the subject, and mediated by the subject's ways of experiencing the world. Thus, the 'object' becomes a transcendental notion, rather than something concrete: 'things' disappear from the human horizon: they become 'an sich'.

Time as we measure it, with the help of a clock, is 'spatialised', and made measurable by awarding numerical value to it. This enables us to think time in terms of numbers, as a series of isolated points, infinitely small moments that follow each other sequentially (Massey 2015). But time for Bergson can only be thought in terms of constant renewal. Lived time is typified as 'the new in the making' (le nouveau en train de se faire). Thus, the sequential view on time as a sequence of instances is flawed, or, at best, a useful instrument within specific approaches of reality. This position bears a striking resemblance to the ideas of another twentieth century philosopher who also attempted a rethinking of time: Martin Heidegger. Heidegger rarely referred to the works of Bergson, but he does do so in the introduction of 'Sein und Zeit' (1962 [1927]; for a discussion of the reasons why, see Massey (2015)). Here, Heidegger portrays him as a thinker who remains locked into Aristotelian thought, without managing to go beyond its limitations (Canales 2016). But Heidegger misses a crucial point in his criticism of Bergson: namely, that time, as durée (living experience, lived experience, evolving life etc.) is never to be thought in spatial metaphors. Heidegger states Bergson turns around the Kantian table: thinking space as time, rather than time as space. But the whole endeavour of Bergson is to think time as life, not as space, nor vice versa. Bergson in all his works opposes this conception of time. It is what Bergson criticises in his wording of 'temps' versus 'durée'.

In a note in Being and Time (Sein und Zeit (1927, pp. 432–433), quoted by Massey, (2015), Heidegger also criticised the spatialisation of time. In his criticism he referred to the Aristotelian tendency to think time as ἀριθμὸς κινήσεως [arithmos kineseos] – a numerical succession of moments. Heidegger believes this conception of time to be flawed. In 'The Creative Mind' Bergson, an intuition of duration puts one in contact with multiple forms of duration, which one might follow upwards towards the spiritual or downwards towards matter (2007 [1923] p. 187). In this regard Bergson's view on intuition concerns an immediate contact between subject and object.

Bergson would attempt to overcome the tension between freedom and determinism by thematising time in relation to living processes: as vital impulse. For Bergson, the evolutionary process should be seen as the endurance through time of a creative rather than mechanical force. This creative force is responsible for the continuous generation of new shapes and forms. Vital impulse is the creative force of these continuous processes of divergence and convergence. Because time is a flux of change, and not a sequence of causally related events, there is room for consciousness and freedom as emerging aspects *from within* this flux of time. And it is here that immediate concreteness is to be found.

Whereas Kant needed to develop two ontologies: the deterministic ontology of nature and the autonomous ontology of freedom, Bergson develops an ontology that can encompass both nature and the human subject. And whilst Kant's ontology was basically an attempt to reconcile an ethics of autonomy with the mechanistic worldview of Newton, Bergson's position would inevitably conflict with the latter. For Bergson, both freedom and causation are aspects of the flux of change. The apparent incompatibility of freedom and causation as perceived by Kant was a consequence (a symptom) of the mechanistic worldview. This would have consequences for Bergson's dialogue with the exact sciences of his time.

2.1.2   Vital Impulse

Vital impulse is a second concept central to Bergson's thought. Where other philosophers in the past had taken physics as the master science, and had therefore focused their work on the implications of physics for our worldview, Bergson sought to establish a new metaphysics based on and in relation to the life sciences – at that time revolving around Darwin's theory of evolution. Vital impulse designates a non-mechanistic account of evolutionary processes. An important aspect of the vital impulse theorem is the idea that there is an inherent tendency towards change in organisms that is independent of the environment. Whereas Darwin believed that organisms are forced to change in response to changing or challenging environments, Bergson believed that evolution is an intrinsic part of the flux of change, and that there is an intrinsic impulse towards change at work in living beings. This tendency towards change is already visible in the principle of genetic variation, a process that predominantly occurs between generations rather than during the lifespan of an individual organism. Vital impulse necessarily relates to teleology.

Teleology adheres to the idea that everything exists in a striving for pre-established, potential end states. This thesis, originating in the metaphysics of Aristotle, has a certain plausibility, especially in biology. Evolution seems to move in the directions of increasing complexity and increasing levels of self-awareness.

Aristotle, in his definition of a final cause, presupposed that all things strive for an end, a pre-determined form: hence the acorn becomes an oak tree, and the embryo becomes a human. This telos is pregiven and a causal factor: it is as if we are pulled towards it.

Previous theorists of metaphysics like Hume, Kant or Berkeley were, in their relation to science, often focused on establishing an ontological ground for the determinist views of classical mechanics (Newton), explaining causality, predictability, quantification and the static nature of universal laws. Bergson instead aimed to develop an ontological account of the unpredictable development of complex systems, continuous change, transformation of organic forms in their qualitative nature and the irreversibility of becoming. In his approach to such a radically new metaphysical account, however, Bergson wanted to avoid both a mechanistic and a classical teleological stance. For him, the world is in flux, but not moving in the direction of a pre-established ends: vital impulse in effect is a concrete manifestation of time, understood as the way in which life forms endure and evolve over time, but not with a predefined path towards some type of pre-established, or pre-establishable end state.

Teleology is persistent in public understanding of biology, for even though Darwinian evolutionary theory is founded on the principle of random variation, depictions of, for example, human evolution usually show a clear line of progress from ape to man, with man being the ultimate end of the evolutionary ladder. This is even visible in the left-to-right depiction of this evolutionary ladder, with an increase in length of the various specimen involved. The persistence of this teleological worldview is also visible in interpretations of evolutionary thought that were voiced around 1900- many authors considered evolution as the gradual realisation of a pre-existing ideal, a manifestation of a pregiven fate. In other words, the process of evolution seemed destined to realise a divine purpose. And as such, strict divisions between human life and other life, as well as between life and nonlife could still be drawn.

Contrary to the biological vitalists of the nineteenth century,[8] Bergson's vital impulse does not entail a strict delimitation between the abiotic, inorganic world, and the living, organic, or in Bergson's words, organised world: it undermines the idea, at that time common, that the two are separate. It signifies a further breach, though. Nineteenth century vitalism held that living organisms are *essentially* – in their ontology, their very being "different from non-living entities because they contain some non-physical element or are governed by different principles than are inanimate things" (Bechtel and Richardson 1998). Whilst referring to his own philosophy as vitalist, Bergson sees this 'non-physical' aspect of life as a characteristic of the process-nature of reality – of time itself, in terms of duration, rather than as some kind of metaphysical, external principle. Although defining élan vital as the

---

[8] E.g. Jöns Jakob Jansz Berzelius, Hans Driesch, Johannes Peter Müller but also Louis Pasteur, in specific in relation to his experiments on fermentation. Its primary view being that life is not fully determined by laws of chemistry and (ultimately) physics (for a more elaborate account, see Chap. 3).

original common impulse that explains the creation of all living species, it is an emergent property of time itself rather than a mystical life force. It is the source of the multiple and evolving nature of reality, both in the biotic and in the abiotic realm.

Apart from genetic variation, Bergson also discerned a psychological factor at work in evolutionary processes: an individual tendency towards change. The occurrence of evolution is not dependent of mere haphazard randomisation; it is first and foremost dependent on an inherent strive for change (Caeymeaux 2013, p. 53). This strive for change should not be confused with individual and subjective human psychological intentionality. It might best be compared with Nietzsche's and Schopenhauer's concept of the Will. Vital impulse thus refers to the apparent tendency towards self-organisation in living systems.

In *Bergsonisme* (Deleuze 1966) Gilles Deleuze explains vital impulse as an internal force in which the distinction between organic living structures – or better, processes – on the one hand, and inorganic (and organic non-living) systems and processes on the other is indiscernible. In that case, the emergence of life is no longer something that has been triggered by something external. Life should rather be seen as an internally emerging property in complex systems.[9] Ultimately, Bergson's concept of vital impulse is more comprehensive than the nineteenth century biological vitalist notion of a spark of life. With vital impulse, Bergson describes the process-nature of reality. It therefore incorporates a specific view on temporality. As Bergson puts it, biological vitalism is "a sort of label affixed to our ignorance" concerning the nature of life. Any mechanistic reduction of life "invites us to ignore that ignorance" (*Creative Evolution*, pp. 42). So, essentially Bergson does not seek to define vital impulse in vitalist terms, as an external factor that triggered the origin of life, and continues to characterise it. His notion of vital impulse rather seeks to describe complexity, something which we now understand as the self-emerging nature of living systems and processes. Thus, vital impulse can be regarded as a predecessor of complex systems theory and chaos theory; it does not embrace the idea of a 'spark of life' that has elevated us (sponges, fungi, plants, animals, humans) above base, abiotic matter: life emerged out of itself. As such, Bergson's philosophy cannot be equated with the vitalist theories of nineteenth century biology, although unfortunately this does occur quite often. I will discuss this point more elaborately in paragraph 6.3.

According to Bor (1990), vital impulse, as the driving force for change, is identical to duration. They both designate time as manifest in the ever-changing and self-revealing nature of reality. Vital impulse is manifestly present in the givenness of conscious experience. But whilst it is clear that vital impulse diverges from the vitalist movement in nineteenth century biology, it merits further concrete definition. For Bergson vital impulse explains not how evolution of organisms can occur, but rather what evolution is, how organisms develop.

---

[9] Elie Metchnikoff, a leading Russian immunologist and contemporary to Bergson, specified the intrinsic nature of this emergence in terms of the organisms ability to self-define (see Chap. 4).

Bergson's main criticism of mechanistic determinism is its presumption that each new development is already contained in the preceding ones. In this reduction there is no place for renewal. And as a result, mechanistic reductionism cannot account for change. His alternative is a 'reversed teleology' that outlines a better understanding of the notion: the final cause of life is to be found at its beginning, rather than its end. It does not predeterme an 'end state', since this would exclude the radically new. This repositioning of teleology bears several similarities to the notion of negentropy, a term coined by the physicist Léon Brillouin (1889–1969) who defined it as the ability to retain information (Brillouin 1956).

To explain the notion of negentropy, we have to turn to the earlier phases of thermodynamics. In the preliminary studies that would lead to Thomson and Kelvin's definition of the second law of thermodynamics (Thomson 1851), Sadi Carnot and Rudolf Clausius described a general tendency of the interrelation between energy differences and mass. The main importance of this second law is that it shows how isolated systems will always tend to an increase in entropy (Carnot 1824; Clausius 1850). thermodynamics informs us that the universe in general tends to distribute the differences in organisational structures as evenly as possible: entropy hereby defines the measure of chaos, in contrast to ordered systems. This means that systems cannot spontaneously increase their order, necessitating an external influence. And introducing such an external influence will inevitably decrease order elsewhere.

The principle of negentropy, or self-organisation, contradicts this. Erwin Schrödinger's definition of life (Schrödinger 1944) derives from negentropy: Life's self-organising properties go against the second thermodynamic law. Negentropy also accounts for why life arrives at ever greater levels of complexity without any external organising influence.

Although only confirmed by chaos theory in the latter part of the twentieth century, Bergson's view of life as a self-organising creative principle preludes these views. In this sense, vital impulse should not merely be regarded as something restricted to what is alive, as some kind of property that is impressed on inorganic matter from the outside. It should rather be seen as an intrinsic property of all complex systems and the processes in which these systems manifest themselves.

As said, several associations are possible from Bergson's vital impulse to other strands of, predominantly German, metaphysics of 'will'. It might be compared to Leibniz's philosophy, as well as that of Schelling. Vital impulse might be compared to Schopenhauer's concept of 'will', more specifically: the 'will to live' (German: 'Wille zum Leben'). Nietzsche's concept of 'will to power' (German: 'Wille zur Macht'), as inspired by Schelling,. Besides these more obvious parallels with Leibniz, Schelling, Schopenhauer and Nietzsche, some parallels can also be discerned between the concept of vital impulse and the Darwinian-Mathusian notion of struggle for existence. Here, however, an important distinction should be drawn: where the notion of a struggle for existence presupposes an intrinsic property in life forms that makes them want to exist and strive for survival ('continuation'), vital impulse describes merely a tendency: complex processes self-organise, diverge and individuate, but merely because they have evolved to do so, not on the basis of some kind of intended striving.

In Bergson's notion of vital impulse, no teleology in the classical sense is involved, merely the properties of the molecules involved in the life process; they do not derive from notions of will, ends, desires etc. Vital impulse permeates everything, from molecular interactions via linguistic structures up to the structures of the universe. It is that which might be referred to as the driving force manifesting itself in everything as the 'real'. Thus, vital impulse, is the way in which time is expressed in matter (Bor 1990). Bergson defines matter as follows: "Matter, in our view, is an aggregate of 'images'. And by 'image' we mean a certain existence, which is more than that which the idealist calls a representation, but less than that which the realist calls a thing – an existence placed half-way between the 'thing' and the 'representation'. Matter, in other words remains something that exists between perception and reality, rather than being the 'substance' of reality: substance as such is not the *prima materia*: duration is, whilst it is not material in nature. Matter is part of what Bergson designates as the 'virtual'. Here, the virtual does not refer to some type of hyper-reality but to the sphere of becoming, It is suspended between observer and observed.

2.1.3   Duration

Duration (French: *durée*) is "the continuous progress of the past which gnaws into the future, and which swells as it advances" (*Creative Evolution* pp. 4). It is contrasted with objective, quantified time (French: 'temps'), as Bergson uses this term in his 1922 debate with Einstein (see Chap. 3). At that time, most participants in the debate saw duration merely as subjective, psychologically experienced time, whilst 'temps' was associated with objective time. Einstein likewise tended to regard Bergson's concept of duration as merely applicable to the psyche, whilst to his mind 'temps', as 'that which is measured by clocks', held a supra-subjective validity. Duration, in Einstein's view, referred to the individual, subjective experience of time whilst temps refers to mathematical, objectively measurable clock time. This concept of measurable time understands time as if it is a spatial category. As bodies in three-dimensional space can be ordered in sequence, next to each other, but can never occupy the same spot, moments in time will always follow the one after the other, and never occur simultaneously. The major flaw of the latter conception is that it negates the continuous fluidity of change. Bergson's observation concerning the interconnection between clock time and space is that, although it makes time measurable and calculable, it obfuscates time as something real, something which endures as continuous change. Time, as measurable unity, is thus an artificial reification of duration, transforming it into something spatial.

Bergson's criticism of our conventional conceptualisation of time was that it fails to account for the way in which we and other life forms undergo and experience duration. To illustrate this notion of 'lived time': I was, enthusiastically typing away a – for my possible reader – what still amounted to a somewhat technical understanding of Bergson. At the time I wrote:

"I am not aware of the fact that it is now already 00.43h, on the 26th of October 2016. But I am aware of my thoughts, the visit shortly before of my neighbour still persists in my memory. And both the wood of the rim of my dining table and the pauses I take in inhaling my hopefully last cigarettes are things that endure in this flowing now. The flowing now consists of both me and the things that I experience: but, more directly, my immediate experience is made up of this experience: it constitutes both the notion of cigarette, neighbour and me. My grandmother was buried more or less exactly a year and 10 months ago, and this temporary factum – whilst thinking of her – is not relevant to my writing. I also notice the feel of the plastic of the keys on my keyboard, the smell of the smoke drifting around in my room, the sound of the instability of my laptop that lacks one of its four rubber stands on the bottom, and the background noise of the cement factory, hollowing out the other side of the shrinking hill that I would be able to look upon if my curtains weren't closed, as well as the weird framing of my round glasses. This paragraph itself however, is not something that makes up immediate experience. And in a sense, this written text does not form part of immediate experience (although writing it does, and so does your reading of it, and any sensory perception that guides it)".[10]

In this regard, immediate experience is the locus for duration. It is not only experienced by human subject. It is experienced by all that is alive (see Chap. 4).

With the concept of duration, Bergson hoped to break with the traditional subject-object divide that was such an unalienable part of Western metaphysics since René Descartes (1596–1650) and Immanuel Kant (1724–1804). Kant believed that we can only know the world as it *appears* to us, not as it is in itself, referring to the latter as the *thing in itself* ('Ding an sich'), sometimes equated by Kant's notion of the noumenon. For someone with basic understanding of classic Greek, who at the same time keeps in mind the distinction between names and things discussed earlier with regard to the problem of the rose, it may seem counterintuitive to use the word noumenon for 'things in themselves'. 'Noumenon' in classic Greek means "something that is thought".

Our modern distinction between language and reality was not made in classic Greek philosophy. Thus, 'thought', 'name' and the object thereof were the same, as can be exemplified by the notion of 'logos', which counted as the structure of the world, the structure of thinking and the structure of language at the same time. In Kantian philosophy, the notion of a noumenon designates the world of things outside of human experience, that aspect of things that can only be thought, or rather, 'assumed', rather than sensed.

The noumenon is that what exists outside of human experience. The phenomenon how things appear within human experience. For Kant, thought leads us to distinguish between the noumenal and phenomenal. But to attain knowledge, we can only resort to a transcendental realm, the discursive world of arguments and reason, since knowledge cannot have direct access to either the world that hides behind our naming of it or the world that is constituted in our experience. Thus, the noumenal and phenomenal are a states of affairs that can only be deduced by reason. Within the Kantian theory of knowledge, we thus remain bound by the limitations

---

[10]As time goes by, the cement factory has by now ceased its activities, delivering its surroundings of its ominous noise at night. The several square kilometer hollow in the hill is to become 'planned nature', whilst my grandmother's demise will be 5 years ago in 8 days.

of our ways of knowing the world, namely in terms of space and time. Bergson agreed that we may indeed have such limitations, but only for rational analysis, not for intuition. Only in the analytic rational approach, distinctions between subject and object, monism and plurality, particular anpd universal, cause and effect become relevant. For intuitive ways of knowing the world, this is not the case. Furthermore, Bergson is critical of the representation of time as similar to the homogenous environment of space, as we have seen, in which objects supposedly find themselves: as a homogenous environment in which events take place. Time should not be interpreted as a subjective form of human sensibility (Bor 1990, p. 153–154). Instead, time should be seen as duration, something that persists, as a process. This repositioning also affects the Kantian notion of causality.

Duration is a much broader concept than merely the subjective and psychological experience of time. Consider the following question: 'Does the brook flow through the valley because it lies below the hills, or does the valley lie below the hills because the brook flows through it?' Causality here appears as a misguided way to interpret the world, albeit one we are discursively accustomed to. Without the brook, no valley; and no valley without the brook. They are both part of the same event. The question can only be resolved by replacing traditional notions of causality with a different notion of time. It does not help to argue that, while concepts like brooks and valleys are human (all-too-human), interpretations, concepts such as friction and gravity are more 'objective', because these latter concepts are anthropogenic in nature. We humans are the ones who experience our world in terms of valleys and hills, friction and gravity. Like Bergson, Gilles Deleuze criticizes dominant interpretations of the importance of clock time.[11] Time for both is first and foremost lived time. It cannot be isolated from the living structures in which it manifests itself. Bergson considers clock time as merely a metaphorically reduced way of depicting time in terms of space. Again, time becomes represented as a series of moments, measured in temporal millimetres (seconds, or fractions of a second) as it were. Lived time is of a wholly different character.

Only in the here and now do we have an immediate access to a sensory awareness of time. Here, the world and our experience of that world are not separate, and both reveal a process nature. But our understanding of 'here' and 'now' tends to be preconditioned by a deterministic framework, one that takes a position quite similar to a snapshot interpretation of the present. Furthermore, for Bergson, freedom is not a given. Rather, it is our imagination of past and future states that creates the possibility and experience of freedom. In *Creative Evolution*, Bergson describes different levels of this relationship between freedom and determinism. Determinism is symptomatic of seeing something as bound to the present. In his view, the inorganic world is bound to its present states and therefore more determined than the organic living world, or, in his words, the organised world. Animal life has some ability to

---

[11] Although he takes a different point of view towards the metaphor of film: Bergson considered cinema an illustration of the wrongful spatialisation of time whilst Deleuze considers cinema from a much more positive point of view and ultimately even interprets Bergson's philosophy as cinematographic in nature.

function beyond the determinants of its present states – it is therefore not fully determined by external impulses (environment) and internal impulses (instincts). Human life, as a result of memory and imagination, acts upon past states (represented in archives) and envisaged future states, and is therefore even more free. Thus, any concept of freedom is welded with an experience of temporality.

Memory and imagination constitute the basis of our ability to act freely, but they are also responsible for our wrongful conception of time as a line, with the present as a dot moving from past to future. In the conception of time as objective and measurable by clocks, we, according to Bergson, forget that we are not dealing with real 'time', but with our own construction of time. Time is not only measured by a clock, for clock time is also *produced* by instruments. Bergson considers time as studied by historians, for example, as a qualitative approach to time, whilst time as studied by physicists is fundamentally quantitative. As such, Bergson's views on *temps* and *durée* inform us that time as it is studied in physics is a conceptual tool, rather than a physical reality. Therefore, he criticises contemporary physicists, most notably Albert Einstein, of unwittingly interpreting their views on time in metaphysical terms. This did not go well with Einstein. It would trigger his problematic debate with Einstein on time in 1922. As mentioned, I will discuss this debate in more detail in Chap. 3.

Literature

Ansell-Pearson, K. (2018). *Bergson: Thinking beyond the human condition*. New York: Bloomsbury Academic.

Aristotle. (1984). *On sophistical refutations*. Trans. by W.A. Pickard. Cambridge.

Barnard, G. W. (2012). *Living consciousness: The metaphysical vision of Henri Bergson*. Albany: Suny Press.

Bechtel, W., & Richardson, R. C. (1998). Vitalism. In E. Craig (Ed.), *Routledge Encyclopedia of philosophy*. London: Routledge.

Bergson, H. (1908 [1889]). *Essai sur les données immédiates de la conscience*. 6th ed., 1908, WS, IA. Doctoral dissertation.

Bergson, H. (2001 [1889]). *Time and free will: An essay on the immediate data of consciousness*. Trans. F.L. Pogson, London: Macmillan, 1910, IA, HTML; New York: Harper & Brothers, 1960, IA; Dover, 2001.

Bergson, H. (1988 [1896]). *Matter and memory*. Translated bys N. M. Paul & W. S. Palmer. New York: Zone Books.

Bergson, H. (1912 [1903]). *An introduction to metaphysics* (trans; Hulme, T.E.). Hackett Publishing Company 1999. Also included in: Bergson, H. (2007 (1923)). *The creative mind: An introduction to metaphysics* 1923. Dover Publications: Dover.

Bergson, H. (1911 [1907]). *Creative evolution*. New York: Henry Holt and Company.

Bergson, H. (1927a). Banquet speech. In H. Frenz (Ed.), *Nobel Lectures, Literature 1901–1967*. Amsterdam: Elsevier Publishing Company.

Bergson, H. (1927b). Lettre à Léon Brunschvicg, 22 février 1927. *Journal des Débats*, 28 février 1927.

Bergson, H. (2007 [1923]). *The creative mind: An introduction to metaphysics*. Dover: Dover Publications.

Bor, J. (1990). *Bergson en de onmiddellijke ervaring*. Meppel: Boom.

Brillouin, L. ([1956, 1962] 2004). *Science and information theory*. Mineola, NY: Dover.

Brougham, R. L. (1993). Ontological hermeneutics: An overlooked Bergsonian perspective. *Process Studies, 22*(1), 37–41.

Buford, T., & Oliver, H. (2002). *Personalism revisited: Its proponents and critics*. Amsterdam/New York: Editions Rodopi.

Caeymeaux. (2013). The comprehensive meaning of life in Bergson. In S. M. Campbell & P. W. Bruno (Eds.), *The science, politics, and ontology of life-philosophy* (p. 2013). London: Bloomsbury.

Canales, J. (2016). Einstein's Bergson problem: Communication, consensus and good science. *Boston Studies in the Philosophy and History of Science, 285*, 58.

Čapek, M. (1971). Bergson and modern physics: A re-interpretation and re-evaluation. *Boston Studies in the Philosophy of Science, 7*. D. Reidel Publ. Comp.

Čapek, M. (1978). Bergson, nominalism, and relativity. *The Southwestern Journal of Philosophy, 9*(3), 127–133.

Carnot, S. (1986 [1824]). *Reflections on the motive power of fire*. Manchester: Manchester University Press.

Clausius, R. (1850). Ueber Die Bewegende Kraft Der Wärme Und Die Gesetze, Welche Sich Daraus Für Die Wärmelehre Selbst Ableiten Lassen. *Annalen der Physik, 79*, 368–397, 500–24. Translated into English: Clausius, R. (July 1851). On the moving force of heat, and the laws regarding the nature of heat itself which are deducible therefrom. *London, Edinburgh and Dublin Philosophical Magazine and Journal of Science. 4th, 2*(VIII): 1–21, 102–19.

de Cluny, B. (1906 [1515 – 12th century]). *The scorn of the World: A poem in three books*. (transl. and ed. Preble, H., Jackson, S. M.); *The American Journal of Theology, 10*(1): 72–101; (prologus and book 1), *10*(2): 286–308 (book 2); *10*(3): 496–516 (book 3).

Deleuze, G. (1991 [1966]). *Bergsonism* (trans. Tomlinson, H., Habberjam, B.). New York: Zone Books.

Heidegger, M. (1962 [1927]). *Being and time* (trans. Macquarrie, J., Robinson, E. Oxford: Basil Blackwell.

James, W. (1975 [1907]). Pragmatism: A new name for some old ways of thinking. In F. Burkhardt et al. (Eds.), *The works of William James. Vol. 1*. Cambridge, MA: Harvard University Press.

Kant, I. (1781/1787 [1929]). *The critique of pure reason* (trans: Smith, N. K.). Macmillan, Pub. 1929, Press (A edition 1781 + B edition 1787).

Legge, J. (Trans) (1893). *The Chinese classics*. Oxford: The Clarendon Press.

Leighten, P. (2013). *The liberation of painting. Modernism and anarchism in Avant-guerre Paris*. University of Chicago Press.

Massey, H. (2015). *The origin of time: Heidegger and Bergson*. Albany: Suny Press.

Nehamas, A., Woodruff, P. F. (Trans.). (1995). *Plato: Phaedrus, with introduction and notes*. Indianapolis: Hackett.

Petrov, V. (2013). *Bertrand Russell's criticism of Bergson's views about continuity and discreteness*. Published for the International workshop Philosophical Analysis: Concepts, Meaning, Language, Sofia, 24 October 2013.

Riquier, C. (2016). The intuitive recommencement of metaphysics. *Journal of French and Francophone Philosophy - Revue de la philosophie française et de langue française, XXIV*(2), 62–83. (trans: Beranek, E.).

Russel, B. (1912). The philosophy of Bergson. *The Monist, 22*(1912), 321–347.

Schrödinger, E. (1944). *What is life – The physical aspect of the living cell*. Cambridge: Cambridge University Press.

Sfara, E. (2015). *La philosophie de Georges Canguilhem à travers son enseignement, 1929–1971: examen du concept d action*. Philosophie: Université Paul Valéry – Montpellier III.

Tagore, R. (1915). *Sadhana, book II: The problem of self*. New York: MacMillian.

Thomson, W. (1851). On the Dynamical Theory of Heat, with numerical results deduced from Mr Joule's equivalent of a Thermal Unit, and M. Regnault's Observations on Steam. *Transactions of the Royal Society of Edinburgh. XX (part II)*, 261–268, 289–298.

Vrahimis, A. (2011). Russell's critique of Bergson and the divide between analytic and continental philosophy. *Balkan Journal of Philosophy, 3*(1).

Chapter 3
Time and Life: Bergson and Physics

*The pure present is an ungraspable advance of the past
devouring the future. In truth, all sensation is already memory.*

- Henri Bergson, Matter and Memory

Throughout his oeuvre, Bergson addressed a tenacious problem in the conceptualisation of time in the exact sciences. In his view, approaches in classical mechanics attempted to 'think' the universe outside of time. Classical mechanics did embrace a concept of time, but it was a concept of time as a mechanical process; time reduced to spatiality. With this concept, events are supposed to enfold as if occurring in a linear process. As Bergson would write to his friend and colleague, the American Pragmatist William James:

> "I had remained [...] wholly imbued with mechanistic theories, to which I had been led at an early date by the reading of Herbert Spencer. It was the analysis of the notion of time, as it enters into mechanics and physics, which overturned all my ideas. I saw, to my great astonishment, that scientific time does not endure.[1] that positive science consists essentially in the elimination of duration. This was the point of departure of a series of reflections which brought me, by gradual steps, to reject almost all of what I had hitherto accepted and to change my point of view completely." (letter of Bergson to James (1935 [1908])).

The conception of time as a linear mechanistic process distorted its nature, thus robbing it from its essential property: that it endures. To enable this awkward conceptualisation of time, mechanistic thought reconceptualised time in a vocabulary associated with space. As Bergson biographer Milič Čapek phrased it, the worldview of classical mechanics took a container perspective on space and time (Čapek 1971). From this perspective, space is considered as some kind of container for all extended objects whilst time, imagined in spatial terms, is seen as a container for events. This is a view that we have grown so accustomed to that we take it for

---

[1] Although not sufficiently explicit about the matter, Bergson did acknowledge that this position - that scientific time does not endure - does not mean that only life endures, and nonlife does not. This is an important note, since it is relevant for the enduring discussion whether Bergson's philosophical vitalism follows the strict division between life and nonlife in biological vitalism or not. See the next chapter for a discussion of this point.

© Springer Nature Switzerland AG 2021
L. Landeweerd, *Time, Life & Memory*, Library of Ethics and Applied
Philosophy 38, https://doi.org/10.1007/978-3-030-56853-5_3

granted. Bergson even conceives of it as a natural tendency of how our minds function.

Although clocks are useful instruments to render temporal events calculable, especially in scientific research practices, Bergson warns against edifying clock time as an objective, real category. Bergson's most important objection against identifying time with clock time is that itleads to a representation of time that confuses duration with extension. What is measured by clocks is not real time but constructed time. In contrast, lived time (duration) is time as it becomes manifest in ever-changing and ever-evolving processes in nature. It not only refers to how we experience time, but to living systems in general.

For the natural sciences there are two concepts of time that are to an extent incommensurable (de Saint-Ours 2008): time as a container of events (clock time) and time as becoming (change). These concepts reflect two ways of experiencing time, namely time as discrete (especially in distinguishing between past, present and future) and time as continuous. Bergson defined a way to reconcile these two concepts. (see Chap. 5).

The concept of time as a container was dominant from Newton onwards. It construes time as something absolute, as if it were a container for all events, disregarding their ever-changing evolving nature. 'Becoming', regarded from this perspective, is merely a process unfolding in time, not time itself. Newton represented this container view on time by depicting time as a straight unchanging line. Einstein, radically innovative as he may have been, also retained an absolute conception of time, albeit not in isolation, but in its relation to space. He represented time as a plane that, although plastic, remains understood through the lens of space and determined by its relation to mass and gravity. As such, spacetime remains a geometric notion. Both the notion of linear time (Newton) and the notion of spacetime (Einstein) allow for the idea that time can exist without change. Leibniz already developed a somewhat alternative view on time by seeing time as relative: Leibnizian time and space, in the words of de Saint-Ours, are "[…] just relations. Space is the order of coexistences while time is the order of successions" (de Saint-Ours 2008, p. 3). Yet, this conception of time, although relational, remains caught in classical determinism, since it represents time as a geometric category, a fourth dimension. This is best demonstrated in the notion of a 'block universe' (Petkov 2005): the idea that time is viewed as a 4th dimension. This view, entails that past, present and future state are determined points in a four-dimensional cube, as much as here, there and yonder are places in threedimensional space. 'Passage of time', in the block universe, is an illusion, and change a mere epiphenomenon. As in the classical assumption of threedimensional space an linear constant time, there is no place for *true* randomness if we accept the block universe notion of time as a fourth dimension. Randomness might be a phenomenon to us, but it does not 'really' occur. It becomes reduced to 'what one cannot predict' rather than 'what has not been fixed in advance': it is merely the restrictions of our intellect that create the semblance of randomness whilst reality 'in itself' does not demonstrate randomness.

As mentioned in the previous chapter, Immanuel Kant understood time as part of how our experience of the world is structured. Becoming is something that we perceive as a result of this epistemological substructure. The Kantian notion of time entails an epistemological subjectification of this concept of time. Speaking in Kantian terms, time is a medium that enables our experience of becoming, whilst in Newtonian and Einsteinian terms,[2] time is a parameter of the universe that is either fixed and unchanging (Newton), or plastic but still determined in its relation to speed and gravity (Einstein). As such, one might say that the whole project of the scientific revolution, from its early beginnings around 1500 until the early twentieth century were aimed at an abolition of change.

In the dominant view in physics until quantum mechanics, becoming thus appeared merely as a way in which events unfold, rather than conceiving of clock time as a particular way of ordering this process of becoming. In the exact sciences, the return of 'time as becoming' can be found in evolution theory and thermodynamics (although what is at stake in the latter case is a process of 'degrading' or dissipation, or 'negative becoming', rather than true 'becoming'), and, more recently, in chaos theory. Here, an undertow view on time suddenly challenged the idea of time as a container of events. In its earlier stages, this alternative view was still reconcilable with the dominant view, but ultimately the incompatibility of these two paradigms became apparent and a clash could not be avoided. It would become quite manifest in the debate on time between Einstein and Bergson, but other facets of the issue became apparent in other contexts as well.

The issue to what extent time can still be quantized (whether it can be compartimentalised or should only be regarded as a fluid continuum) remains up for debate. Bergson's point of view would tend to the notion of fluidity, although compartimentalisation, in his perspective, would likely be termed an example of our way of theorising the world,[3] not a property of that world. Here, however, the blurring between spectator and event that is part and parcel of the debates in quantum physics informs an even more radical perspective that might fit better with the, non-definable, immediate experience as discussed by Bergson.

The combined concepts of linear time and three-dimensional space in classical mechanics give rise to a specific view on phenomena such as force, gravity, movement, acceleration: these phenomena occur as determined. Events thus follow each other like a train that can only move forwards, with a continuous speed, and on tracks that never bifurcate. The alternative view on time as endurance *and* change is process-based, complex, undetermined and does not exclude concepts of free will. In the early 1920s, these two concepts of time and how they should be prioritised were to form the basis of what was to become a deeply entrenched debate.

---

[2] Specifically in how Minkowski's notion of spacetime as 4th dimension was taken aboard in the theoretical grounding of his theories.

[3] As a property of how our language informs an interpretation of experience in terms of an experience of the world as the 'sum of all things'.

3.1 The Conceptual Disarray Between Space and Time

Galileo Galilei (1564–1642) was one of the first modern scientists to study time, specifically in relation to the constant movements of pendulums. Galilei also developed the first standards for time (as he did for space). The basic concepts he developed were further elaborated by, amongst others, Isaac Newton (1642–1727). Newtonian time is to be taken as independent of the observer, to progress at a consistent pace independent of location and to be 'measurable' although 'imperceptible'. In classical mechanics time was thus considered to be one of the fundamental 'scalar quantities'. Time was further regarded as absolute. Newton considered it to be comparable to a scalar quantity such as length.

Time, in Newton's mechanics, was thought of as a homogenous environment similar to three-dimensional space. Although this view came to be contested in physics since the early 1900s, it does persist in other exact sciences. When Immanuel Kant (1724–1804) defined his epistemology in line with questions that emerged from the lack of a foundation of Newton's mechanics, it appeared to have led him to a categorical understanding of time in the same terms as space. In *Time and Free Will* Bergson criticises this tendency to think time in the same way as space: instances as sequential dots on a line. The mechanical conception of time was, in his view, a 'non-time'. Since any process becomes predictable there is no renewal, no true evolution in the Bergsonian sense of the word Time is merely the 'track' on which the occurrences are displayed. The elementary problem with this view on time is that once within the confines of time (and we all are, even if the convention of science is to pretend to have an outsider's perspective), we can no longer even perceive the direction or speed of time. Time becomes, in a sense, reduced to a mere variable. The main problem with time thus appears to be our tendency to conceptualise it as we conceptualise space. As van Dongen lucidly states in his 2014 publication on Bergson (van Dongen 2014), what is a clock other than a circular ruler? Conceptualised as such, it remains restricted in the first place to one-dimensional space: linear, and, measurable accordingly.

From Newton onwards, physics started from the idea of a reversibility of the mechanisms of processes through time.[4] Time itself was not considered to be reversible, but the processes occurring through time were. In this view, time was portrayed as a line on which processes starting from a given state and ending in another state might be turned around, which was then supposed to lead to the original state: in short, a symmetry between going forwards or backwards. In classical mechanics, any process described was regarded as reversible according to this principle. Any view on cause and effect also held that effects, when the process was reversed, could act as cause. This assumption is also called 'T-symmetry' or time reversal symmetry (Lebowitz 2008): it concerns the theoretical symmetry of physical laws under the hypothetical reversal of time. This view on processes in physics is best illustrated

---

[4] Some restrictions were defined to this principle. Ludwig Boltzmann for example defined how this was only possible for microphenomena, not for their macro-aggregates.

with again the metaphor of the clockwork mechanism: if one reverses the spin of a clock, the processes involved will also be reversed. T-symmetry fit well within the concept of causality as it arose in modern science.

The concept of causality, also referred to as causation, is abstract. It indicates how one event (the cause) that temporally precedes, is responsible for another (the effect) that follows in time. It only partially builds onto the four causes defined by Aristotle (namely onto the 'efficient cause'). Up to the early twentieth century, this modern conceptualisation of causation necessitated that these objects would physically touch one another. The ancient Greek concept of an 'aether', a kind of proto-material substance existing in the void of space, was part and parcel of this axiom and explained the influence of planetary movement around the sun without strings attached to the planets.[5]

It took until the nineteenth century for T-symmetry, the reversibility of the arrow of time in physics, to be challenged. The second principle of thermodynamics observed aggregate phenomena, on a macro-scale. These showed properties of a different kind: the increase of entropy, the measure of chaos (as defined in the second law of thermodynamics[6]) shows that the universe appears to have a large-scale tendency to strive for a decrease of difference between different mass and energy levels. This principle was described in terms of an irreversibility of the arrow of time in 1927 by the British astronomer Arthur Eddington (1882–1944). As de Saint-Ours states: "[…] with the exception of the second principle of thermodynamics, the equations of classical physics do not give evidence to time's irreversibility, quite the opposite" (de Saint-Ours 2008). But, the principles of thermodynamics describe an accumulative phenomenon; they remain reconcilable with time reversal symmetry, since they only apply to the slow macro-scale processes of the universe.[7]

The hidden basis of the assumption of time symmetry lies in the mechanistic worldview. It is specifically related to the conception of space that serves as a three-dimensional milieu for objects and the conception of time as the accumulation of infinitesimal moments on a line that serves as a milieu for events. Causality then

---

[5] The concept of an aether is a remnant of Aristotle's metaphysics. For him, causes could not manifest over a distance without some kind of medium. A vacuum, in Aristotle's worldview, could not exist. Following his teacher Plato, he therefore postulated the concept of an aether as a fifth element, to explain how objects could influence each other over distance. René Descartes would elaborate this 'aether' in the seventeenth century in terms of vortices influencing each other, thus explaining the material causation of how particles formed material objects. Although these processes were conceived as complicated, they did remain materially determined. And linear clock-time remained a mechanical backdrop for these materially determined, reversible processes.

[6] Rudolf Julius Emanuel Clausius's (1822–1888) definition of the second law of thermodynamics (1854): "Heat can never pass from a colder to a warmer body without some other change, connected therewith, occurring at the same time." Sadi Carnot (1796–1832) expressed a prior version of the same principle in 1822, on the basis of his idealised heat engine. Planck's statement of the second law might be more recognizable and runs as follows: "Every process occurring in nature proceeds in the sense in which the sum of the entropies of all bodies taking part in the process is increased. In the limit, i.e. for reversible processes, the sum of the entropies remains unchanged." (Planck 1987/1903).

[7] In the end, it took until quantum mechanics for this principle to be radically challenged.

serves as the relation between these two. Bergson quotes the mathematician and astronomer Pierre-Simon Laplace (1749–1827) to illustrate this mechanistic view on time and space:

> "We may regard the present state of the universe as the effect of its past and the cause of its future. An intellect which at a certain moment would know all forces that set nature in motion, and all positions of all items of which nature is composed, if this intellect were also vast enough to submit these data to analysis, it would embrace in a single formula the movements of the greatest bodies of the universe and those of the tiniest atom; for such an intellect nothing would be uncertain and the future just like the past would be present before its eyes." (Pierre Simon Laplace as quoted in Marij (2014) as well as Bergson (1911 [1907])).

In *Creative Evolution* Bergson writes: "A superhuman intelligence could calculate, for any moment of time, the position of any point of the system in space. And as there is nothing more in the form of the whole than in the arrangement of its parts, the future forms of the system are theoretically visible in its present configuration." (1911 [1907] p. 8). Reality, in this worldview, consists of components made up of particles, the smallest of which cannot be divided further; these particles are presupposed to hold definite places in space at a given moment in time. This reality of particles situated in space and time can only be grasped in full by an ideal observer. This observer is supposed to be able to perceive this reality, and at the same time interact with it at least to the extent that it is able to be aware of it. This hypothetical observer has also been dubbed 'the demon of Laplace'. Although sometimes compared to God, it is actually the ideal position of the scientific subject itself. Laplace's demon illustrates a meta-paradigm still held in many scientific disciplines, and remains the dominant worldview in the sciences, albeit due to a somewhat one-sided general understanding of modern physics which thinks in terms of causation rather than processes.

In the same book, Bergson criticises the mechanistic worldview with the following argument:

> "[…] though the whole of the past goes into the making of the living being's present moment, does not organic memory press it into the moment immediately before the present, so that the moment immediately before becomes the sole cause of the present one?-To speak thus is to ignore the cardinal difference between concrete time, along which a real system develops, and that abstract time which enters into our speculations on artificial systems. What does it mean, to say that the state of an artificial system depends on what it was at the moment immediately before? There is no instant immediately before another instant; there could not be, any more than there could be one mathematical point touching another." (Bergson 1911 [1907], p. 21)

As Bergson continues, the concrete conception of time refers to time "along which a real system develops". The abstract conception of time is a construction that "enters into our speculations on artificial systems" – systems such as clocks. In other words, the presupposition of clock-time, in its combination with linear causality, leads us to explain events in the world on the basis of a deterministic framework and not the other way around; it is not the empirical evidence of determinism that confirms the existence of clock-time.[8] The classical picture of linear time needed to be revised. This indeed came to be one of the most important topics for twentieth century physics. Where Newtonian physics was regarded as radical due to its assumption of cause and effect without the objects involved touching each other,[9] Einstein's general theory of relativity was considered a revolution in physics since it implies a different image of time than the linear picture. The general theory of relativity implies a different image of time than the line, which is the reason that it is regarded as revolutionary. The theory of relativity reconceptualises time with a geometrical image in mind: although no longer linear, spacetime is now represented as 4th dimension. And although time is regarded as relative to mass and velocity rather than as absolute, these relations are again determined.

3.2 Tinkering with Time: The Debacle with Einstein

In 1922 the physicist Paul Langevin (1872–1946) organised a meeting between Henri Bergson and Albert Einstein for the *Société française de philosophie*. In 1911, Langevin had written a paper for the Fourth International Conference of Philosophy in Bologna and it was this paper that first drew Bergson's attention to the theory of relativity. Langevin attempted to get more public knowledge of the theory of relativity. The meeting between him and Einstein led Bergson to write *Duration and Simultaneity* (*Durée et simultaneité* 1999 [1922]) in the same year. Unfortunately, what was planned as a convivial meeting between two of the most important intellectual minds of the time became to be a heated debate. As Canales writes: "The meeting had been planned as a cordial and scholarly event. It was anything but that. The physicist and the philosopher clashed, each defending opposing, even irreconcilable, ways of understanding time" (Canales 2016, p. 3).

Bergson had not said anything during Einstein's lecture on the previous day, but he was now drawn into the debate due to Éduard Le Roy's mentioning of the equality but distinctness of the two perspectives of philosophy and physics. Bergson responded that although he respected the physicist's work on relativity, his theories did not make philosophical interpretation obsolete. Einstein responded by his infamous comment "The time of the philosophers does not exist" (Canales 2016, p. 19). Einstein wanted to defend his theory of time as relevant to concrete reality, rather

---

[8] This small turn-around is not sufficient: for Bergson, we are even predisposed to think our world in terms of cause and effect, and in geometrical, spatial categories. Here, Bergson implicitly follows David Hume's relativistic account of causality – a habituation of how sensory experience is interpreted in the mind. But in Hume's philosophy the consequence for knowledge is dire: no fundament. In Kant's attempt, causation becomes a transcendental presupposition to arrive at objective knowledge. Of course, the Kantian edifice is built on, but not pre-ordained to be restricted to, such a categorisation of experience.

[9] Until the nineteenth century, some physicists still adhered to the pre-Newtonian view, hoping to solve the problem of gravity through the assumption of an aether, or the assumption of vortices transferring cause to effect over distance.

than as a mere formal mathematical model. He also believed that Bergson's views on time were erroneously confusing subjectively experienced time (psychology) with objective time. Bergson however was well aware of this potential error, and insisted that with the explanations of science, not all was dealt with. He would explain his views on the delimitation between science and philosophy in writing *Durée et simultaneité*. To sketch the context of the debate, some background information might be helpful.

As we have seen, Newton posited that time and space were absolute. They were to be seen as aspects that remained independent of the observer, and as such as parts of objective reality. This position was long regarded as a preferred frame for physics. It still informs our notion of time. We tend to perceive of time as a line, forcing us to see the present as an infinitely small dot on that line. The whole universe is contained in this infinitesimally small dot: the event. In this view, we are already time travellers. The whole universe travels through time. At a continuous pace, second to second, minute to minute, we travel from past to future. But whilst time was considered to be absolute, constant and unidirectional, physical processes were considered to be reversible.[10] This view of an objective reality defined by the absolutes of space and time, in which everything is causally determined, did however raise some issues concerning the status of knowledge. How could an observer acquire objective knowledge if this observer is determined him−/herself? How can one account for the validity of empirical observation? For philosophers after Newton, it became necessary to reconsider the status of space and time. Immanuel Kant attempted to safeguard the ontological validity of Newton's mechanics by providing it with a metaphysical basis. To do so, Kant explained space and time as part of how we necessarily experience reality, rather than as absolute features of reality itself.[11] The general and the special theory of relativity are considered as a radical breach with this principle in the history of physics.

Einstein did not understand time to merely be a continuous line. Depending on speed and gravity, time can either go faster or slower in different situations. His radical adjustment of Newtonian physics consisted of envisaging time as irregular over space, thus shattering the principle of constant and absolute time. Instead, he posited, time elapses differently for different observers under specific conditions. The time elapsed for an observer moving at a high pace would be shorter than the time elapsed for a stationary observer. Similar variability in the time which elapses from the point of view of different observers occurs under the influence of gravitational mass. In other words, under the influence of differences in speed and mass (gravitation), time for one object is different from time for another. This phenomenon, also referred to as 'time dilation', can only be accounted for by regarding space and time as part of the same continuum. Einstein elaborated this view in his general

---

[10] With the exception of the second law of thermodynamics.

[11] But the idea of absolute space and time was not universally accepted. Gottfried Leibniz strongly criticised absolute space on the basis of its logical inconsistencies. As he put it: if one would move the whole content of the universe one meter to the right – when space would be absolute – nothing would change, although everything would have moved.

theory of relativity. On this specific point of the reimagining of time as a geometrical, dimensional category, Einstein's own (implicitly metaphysical) interpretation of relativity would raise Bergson's criticism. Revolutionary as Einstein's shift of paradigm was, his position still does not release us from the metaphor of geometry.[12]

During the 1922 debate, Einstein made his position on the notion of time in philosophy quite clear. For Einstein, time was either what clocks measured, or it was nothing at all (Canales 2016), implying that the rest was mere subjective, psychological, experience. The main trigger for Bergson's opposition to Einstein's account of relativity was that Einstein continued to define time and temporal phenomena such as simultaneity in terms of clocks, whilst Bergson did not believe that the nature of time could be informed by such a mechanism. "When our eyes follow on the face of a clock, the movement of the needle that corresponds to the oscillations of the pendulum, I do not measure duration, as one would think; I simply count simultaneities, which is quite different." (Bergson, *Time and Free Will*, 1889). Time is not measured through the clock itself, these measurements always need to be related to some external referent, or point of observation (Canales 2016).

At first it appeared that Bergson had presented a strong argument against Einstein during the 1922 debate with Einstein: Einstein supposedly could not grasp the nature of Bergson's position, whilst Bergson, supposedly, could grasp the nature of Einstein's position. Canales even claims that this encounter motivated the Nobel Prize Committee to award Einstein the 1921 Nobel Prize, albeit not for his work on relativity, but for his work on photoelectric effect (Canales 2016).

The 1921 Nobel Prize was deferred since much controversy arose over the nomination of Einstein. The Nobel Committee hesitated awarding him with the prize for physics since relativity went against several basic intuitions and axioms in physics. The controversy officially revolved around the theoretical nature of relativity. Ernst Gehrcke and Philipp Lenard opposed awarding Einstein the prize for his work on relativity. They were critical of the mathematics involved. In the end, Einstein would receive the deferred 1921 prize in 1922, and not for relativity but for his explanation of the photoelectric effect. Although this research would prove to be important for quantum mechanics, relativity, at that time, was considered much more important. The Nobel Prize Committee stated: "[…] the Royal Academy of Sciences has decided to award you last year's Nobel Prize for physics, in consideration of your work in theoretical physics and in particular your discovery of the law of the photoelectric effect, but without taking into account the value which will be accorded to your relativity and gravitation theories after these are confirmed in the future." (Bishop 2003). Einstein, obviously annoyed by the whole affair, would mainly discuss relativity in his acceptance speech.

---

[12] Time, first viewed as continuous and unidirectional is now dependent on change. But determinism remains contingent with Einstein's view on the relative discontinuity of the elapse of time that was described by Einstein. What observers make of the changes they observe and the relationship of those changes with other changes can be dealt with on a situation by situation basis – but *observed* changes do not reveal the *unobserved relation between* changes taking place.

Both Germany's rising anti-Semitic sentiments (Lenard would later support the Nazi party) and Bergson's debate with Einstein seem to have influenced the committee's initial indecision. But in later years and specifically after Bergson's demise, many scholars believed that Bergson's criticism of Einstein's conceptualisation of time as relative, and as deeply intertwined with space, was flawed. Resistance against Bergson's criticism however appears to have stemmed from the circumstantial association of Bergson with anti-scientific mysticism rather than from his arguments proper.

Lettevall et al. (2012) and Canales (2005, 2016) point out that critics often refer to a remark taken up in the appendix of *Duration and Simultaneity* (1999 [1922]) to prove Bergson's apparent ignorance over the scientific aspects of Einstein's general theory of relativity (Lettevall et al. 2012, p. 253). Bergson, in appendix III, discussed the famous travelling clocks experiment (nowadays more known as the twin paradox). In the translation provided by Lettevall et al.: "once re-entering [the earth], it[13] marks the same time as the other." (Bergson 1999 [1922]). Another translation, by Leon Jacobson (Bergson 1965), reads: "immediately upon re-entering, it [the moving clock] points to the same time as the other". However, the full quote, in the translation of Leon Jacobson, reads:

> "In short, there is nothing to change in the mathematical expression of the theory of relativity. But physics would render a service to philosophy by giving up certain ways of speaking which lead the philosopher into error, *and which risk fooling the physicist himself regarding the metaphysical implications of his views*.[14] For example, we are told above that 'if two identical, synchronized clocks are at the same spot in the system of reference, if we shift one very rapidly and then bring it back again next to the other at the end of time t (the time
>
> of the system), it will lag behind the other by $t - \int_0^t \alpha \, dt$.[15] In reality we should say that the
>
> moving clock exhibits this slowing at the precise instant at which it touches, still moving, the motionless system and is about to re-enter it. But, immediately upon re-entering, it points to the same *time* as the other (it goes without saying that the two instants are practically indistinguishable)." (Bergson 1922, appendix III, my italics)[16]

The whole conundrum revolves around particularly the phrasing of "But, immediately upon re-entering, it points to the same time as the other", or, in French "Mais, aussitôt rentrée, elle marque la même heure que l'autre". If we read the English

---

[13] The clock that had been moving.

[14] Bergson did not object to the theory of relativity, but rather to wrongful philosophical extrapolations from it. The following is the French quote: "En somme, il n'y a rien à changer à l'expression mathématique de la théorie de la Relativité. Mais la physique rendrait service à la philosophie en abandonnant certaines manières de parler qui induisent le philosophe en erreur, et qui risquent de tromper le physicien lui-même sur la portée métaphysique de ses vues." In other words: the mathematical expression of relativity is not contested by Bergson, but wrongful philosophical ways of interpreting relativity might confuse the physicist himself in his understanding of relativity.

[15] The alpha in the equation refers to time dilation.

[16] Bergson did not want *Duration and Simultaneity* to have a 7th edition since he deemed his knowledge of the mathematics involved in the further development of relativity beyond his grasp, but since this was not mentioned in his will, further editions were published after his death.

superficially, and the quote isolated from its context, Bergson seems to imply he believed that upon re-entering, the indicators of the traveling clock will shift to the same position as the other clock. The English phrasing 'points to the same time' is as misleading as the original French. If paraphrased in terms of the twin paradox: the travelling brother whose aging went slower does not all of a sudden sprout grey hairs and undergo a wrinkling of the skin upon slowing down and meeting his aged sibling. The English word 'time' however, does not fully cover the same meaning as the original French word 'heure'. Time and 'temps' are more synonymous. 'Heure' means something like 'moment' as much as it does 'hour'. What is meant here is that the moving clock, when re-entering, will start marking the same moments as the other. In other words, the moving clock does not retain its different temporal system when re-entering.

When one reads the full quote, it becomes quite clear that Bergson's understanding of time dilation is correct. He refers to the fact that for both clocks 'experienced time' remains the same in nature, and the 'slowing down' is only noticeable at the instance of re-entering the motionless system. Then, the formerly moving clock again takes the speed of the stationary clock. "The same time" in this last sentence thus refers to the speed at which both clocks work at the re-entering of the moving clock. In this third appendix, both in the full paragraph of the fragment quoted above and in further parts of this third appendix, Bergson quite clearly explains his understanding of, and agreement with the scientific implications of Einstein's relativity.[17] Unfortunately, many readers failed to grasp this wider context. Bergson did not aim to criticise Einstein's scientific claims, but rather Einstein's philosophical inferences from these claims. Einstein's view of time was revolutionary, but according to Bergson he failed to make the decisive step of detaching itself completely from the view of time of classical physics, a view that is mediated by mechanistic thought since it remains tied up with the seventeenth century invention by Huygens of the pendulum clock. Bergson does not deny the value of Einstein's discovery, but

---

[17] Jacobsen's full translation of the paragraph in question: "In a system in uniform translation – the earth, for example, because its acceleration is slight – two identical, synchronized clocks are at the same spot. We shift one very rapidly and bring it back again close to the other at the end of time t (the time of the system); it is found to be behind the other clock by f- adt; if its acceleration was instantaneous at departure as upon arrival and its speed has remained constant, the slowing amounts to t(l-a). No one could express himself with greater precision. Moreover, from the physico-mathematical standpoint, the argument is irreproachable: the physicist ranks the measurements actually made in one system with those which, from this system, appear as if actually made in another. It is out of these two kinds of measurement, merged in the same treatment, that he constructs a scientific world-view; and, as he must treat them in the same way, he gives them the same meaning. Quite different is the philosopher's role. In a general way, he wants to distinguish the real from the symbolic; more exactly and more particularly, for him, the question here is to determine which is the time lived or capable of being lived, the time actually computed, and which is the time merely imagined, the time which would vanish at the very instant that a flesh-and-blood observer would betake himself to the spot in order to compute it in actuality. From this new point of view, comparing only the real with the real, or else, the imagined with the imagined, we see complete reciprocity reappearing, there where acceleration seemed to have brought on asymmetry." (Bergson 1999 [1922], p. 180).

argues that Einstein fails to fully comprehend the newness of his account, precisely because he still continues to define time in terms of clock time, rather than as a fluid process. Bergson in principle embraces relativity theory, but wants to purge it from this atavism: Einstein's tendency to cling to the misguiding (classical, mechanistic) understanding of time as clock time. And whilst Einstein, in the debate, attempted to defend his theories as being more than mere abstract, mathematical conjectures, Bergson actually provided Einstein with the very metaphysics his theories deserved.

The main point Bergson makes against Einstein criticises his metaphysics, not his physics and thus not the theory of relativity proper. In other words, whereas Einstein's physics is post-classical, his metaphysics (his identification of time with clock-time) is still remarkably classical, and Bergson challenges Einstein to make this final step, replacing clock-time by process-time (duration). Bergson's argument with regard to the twin paradox is that 'being present' is not relative: it is not dependent on an external and objective framework of time-reference or for that matter of spacetime-reference. Present-ness is, in the words of theologian Lane Craig "possessed absolutely" (Lane Craig 2016: in spite of his relative obscurity, Lane Craig's reading of the discussion Bergson-Einstein is quite enlightening).

Einstein's main observation in the general theory of relativity was that gravity, matter, energy, and momentum can curve 'flat' spacetime. Several scholars (including Hilbert (1917) and Weyl (1918)) considered the general theory of relativity as a step towards a geometrisation of physics. The problem Bergson saw with Einstein's general theory of relativity was of a metaphysical rather than a scientific nature: one may consider time as separate from threedimensional space (as was the case in classical mechanics), or as part of it (as is implied in the notion of spacetime), but either way, these are ontological starting positions, postulates, rather than scientific outcomes.

Einstein's views on his own theories went directly against Bergson's criticism of the Kantian epistemological tendency to 'think time as space'. According to Bergson, Einstein's implicit philosophical claims remained restricted to the idea of time as a geometrical category: time as a fourth dimension. As such, Bergson was critical of Einstein's reduction of time to a geometric category, and the implied negation of change in the true sense of the word. For within the notion of spacetime, change is merely an illusion, since past, present and future all have their fixed position in the four dimensions (although they appear relative if perceived from within threedimensional space). The theory thus remains fully deterministic, whilst change, in Bergson's view can never be considered change in the true sense of the word if there is no possibility for anything radically new, if every new state is precluded in preceding states.

Bergson considered Einstein's discussion of relativity theory to be "a metaphysics grafted upon science [...] not science" (Bergson 1999 [1922]). As said, during the notorious 1922 debate in Paris, Einstein himself replied that there was no such thing as the concept of time as brought forward by Bergson (Canales 2016, p. 6). There was only the time of physics and the time of psychology. In other words, Einstein believed there was something like subjective time, or time as we experience it, but regarded it as a mere mental phenomenon. Einstein considered this psychological account of time as similar to the time concept of philosophers, and conceived of it as fundamentally psychological: the mental experience of time, without any bearing on the physical world. Bergson's concept of time is misrepresented by Einstein. What Bergson proposes is to replace the equation time = clock-time by the equation time = change, in terms of becoming, perpetual renewal, be it in our experience or in events occurring outside of our minds. Bergson dismisses compartmentalisations between the external and the mental (as will also become evident in his discussion of memory (see Chap. 5)). Bergson considered lived time to be the primary form of time, and did not regard it as mere psychology. For him, lived time encompassed all living systems, including human consciousness. And in stark contrast to Einstein, for Bergson the concept of time as it is used by the physicist was a construction, a secondary phenomenon, compared to lived time as experienced immediately by us.

Einstein tried to eliminate Newtonian time, but his acceptance of the conception of spacetime continues to lean on a Newtonian conception of time as quasi-spatial. In a letter Einstein wrote to his colleague Duncan Haldane, he mentions having received Bergson's *Duration and Simultaneity* and having partially read it aboard a ship headed for Japan. He stated, however, that he had not yet been able to make up his mind over the publication (Canales 2016). At the time, Einstein concluded that Bergson still failed to grasp the distinction between psychological and physical time, although he would ultimately change his position (see end of paragraph in Sect. 3.3). So, here we have two thinkers on time, Bergson and Einstein, who accuse each other in the politest words that the other mistakenly holds categories for real that are in reality merely a part of one's psychological precondition or merely a part of a set of epistemological presuppositions. But which of the two's accusations is legitimate?

Einstein accused Bergson of adhering to the problematic dualistic philosophy of Descartes.[18] But it was actually Einstein who appears to embrace the materialist determinist concepts that are such an unalienable part of Cartesian philosophy (Holton 1973),[19] although he felt more at home in the philosophy of Spinoza than the philosophy of Descartes. It is mainly the term 'relativity' that is to blame for this common misconception: whilst relativism in the humanities implies that truth is relative to context and perspective, Einstein's relativism refers to the irregularity and non-simultaneity of events in time and space. The representation of time as a fourth dimension has become so well known amongst the general public that we

---

[18]A dualism between mind and matter, posited as two separate substances, the one being present in pure form in God, the other in base matter, the human suspended between the two, existing in soul and mind on the one hand, existing in bodily shape like animals, plants and inorganic things on the other: the interrelation between the two would prove to be the most problematic aspect. And although many philosophers after Descartes attempted to find alternatives, the implicit consequences of their worldviews would revert to either the same dualistic paradoxes, or some type of isolated idealism (ultimately digressing in self-defying solipsism (as Schopenhauer discusses in paragraph 2 of book 1 of *The World as Will and Representation*) or mechanistically determined materialism (digressing in a paradoxical position (see last paragraph of Chap. 4)).

[19]Although in some senses, Einstein's philosophical views, as avid reader of Spinoza (Paty 1986) bear more similarities to Spinoza's determinism.

tend to forget that it was derived from the geometrical representation of spacetime as it was defined in 1907 by Hermann Minkowski (Einstein's former teacher) in his interpretation of special relativity. The relation between time, mass and velocity however remains absolute, and, as a result, Einstein's worldview remains deterministic.

Both Einstein and Bergson have made an important contribution to our understanding of time, but on two very different levels. Einstein saw the concept of duration presented by Bergson as an empty signifier, since it lacks a relation to something measurable in the physical world. Bergson however questions the technique of measuring as such as a means to understand time. He does so from a position that is quite similar to William James's: a position that necessarily relies on introspection rather than measurable observation. Einstein's theory of general relativity remains without a doubt one of the most important feats of twentieth century physics. But the concept of time as it is presented in his theory still relies on thought experiments which start from a conception of time as clock time. Due to the presupposition of time as clock time, theoretical physics remained caught up in the logical conundrum over time and space.

During that troublesome encounter in the year 1922, there were aspects of Einstein's theory that Bergson did not understand and aspects of Bergson's conception of time that Einstein did not understand (Canales 2016). Bergson's view that in direct experience, neither time nor spatiality are directly experienced, and only change is might easily be interpreted as referring to the subjective psychological experience of time, but this is not what Bergson intended. Therefore, the debate between the two was almost doomed from the outset. But where Bergson endeavoured to understand the scientific and mathematical principles of Einstein's theory Einstein did not venture into a serious consideration of Bergson's reflections.

Like Bergson, Einstein did not accept Newton's view on time as something independent of what is 'in' the universe. But Einstein's failure to understand and appreciate Bergson's conception of time lies in the fact that Einstein implicitly presumed that his 'time' was a real phenomenon that was independent of the observer and had a physical plasticity: it did not behave as a constant. Here, the scientist seems to be guilty of – quoting William James (also see footnote 25) – the "human trick of turning names into things" (James 1907 [1975], p. 46). Einstein reveals himself to be a 'conceptual realist' rather than a nominalist (see Chap. 2). Bergson can hardly be termed nominalist (Čapek 1971). His views do adhere to the reality of the concrete and the immediately experienced rather than the reality of the conceived. Instead, he took a middle position that might be termed conceptualist: all particular things are understood through concepts, but these lie in the mind, not on some eternal metaphysical plane.

Unfortunately, many still perceive of Bergson's criticism of Einstein as deeply flawed. But the views on time that were embraced by Bergson address a different problem area than the problem area addressed by Einstein. Einstein was preoccupied with the nature of time as relevant to physics, whilst Bergson was preoccupied with the nature of time as relevant to *meta*physics. Einstein's theories, on the level of the underlying mathematics are reconcilable with Bergson's views, and could have formed a basis for a deeper understanding if the two scholars had been able to understand each other's vocabularies. As said, Bergson did not seek to challenge the scientific claims of the special and general theory of relativity. He did not even contest the idea that time could pass differently between two different events measured by two observers either moving relative to each other or differently situated from a gravitational mass. What Bergson did challenge was the way in which Einstein interpreted these claims and ideas. For Einstein " there is no philosopher's time; there is only a psychological time different from the time of the physicist" (Latour 2011). But it appears that Einstein's conception of the reality of 'existence' was what was actually up for debate. The interpretative choices already made by Einstein when entering into his research were the very choices investigated by Bergson. As a result, the debate faltered, which was unfortunate, because it resulted in a growing estrangement between science and metaphysics.

The debate between Bergson and Einstein eventually led to a marginalisation of Bergson's oeuvre. Einstein would eventually, also amongst philosophers, come out as the stronger candidate with regard to the issue of time. The clash between the two minds led to a schism between the worldviews of the sciences and the humanities that is felt to this day. Bergson's criticism of Einstein's special and general theory of relativity seemed to evaporate during the decades after 1922, as did the fame which his work enjoyed during his lifetime. But there were others who took up the same criticism.

Bergson's philosophy was paralleled by other thinkers of the era who also focused on process thinking. One of the most prominent representatives is Alfred North Whitehead. Whitehead's early career focused primarily on mathematics, logic, and physics. Later, Whitehead developed a process philosophy of science. Similar to how Bergson criticised the mechanistic worldview for freezing reality in time, Whitehead criticised the notion (still strong at the time in science) that the world consisted of bits of matter that exist independently of one another. His *Process and Reality* (1929) seeks to flesh out the view that reality should be viewed as a collection of interrelated processes rather than of discrete material objects. His view of reality as a continuous process of becoming avoids the fallacy of misplaced concreteness: the fallacy of confusing abstract concepts for real things.

Whitehead embraced the view that processes are best defined by their relations with other processes. As Bergson was inspired by evolution theory, Whitehead was inspired by Bohr's atomic model (1913) and de Broglies wave theory (1924), whilst the latter's work in general was again inspired by Bergson's philosophy. In a terminology that is strikingly similar to that of Bergson, Whitehead said: "Mathematical physics presumes in the first place an electromagnetic field of activity pervading space and time. The laws which condition this field are nothing else than the conditions observed by the general activity of the flux of the world, as it individualises itself in the events" (Whitehead 1960 [1925]), p. 190)). The demechanisation of the worldview also entails a dethingification of the worldview. Whitehead criticised the notion (still strong at the time in science) that the world consisted of bits of matter that exist independently of one another. Here, physics seems to be troubled by an age-old tendency: the insistence on the idea that the world is a collection of things.

This view is implicitly informed by a traditional atomistic-philosophical conceptualisation of reality as ultimately consisting of material particles. And hardly any physicist today would think of matter as a bunch of inert billiard-ball-like particles. At the quantum level, it appears that, depending on the type of observation one adopts, one may observe either a wave or a particle – whilst these two observations appear to be logically inconsistent. In some views, this forces us to accept the conclusion that reality adjusts itself to the observer. This counterintuitive phenomenon appears as a logical paradox within the classical worldview of a particle-based physics with a linear determinist conception of time. But as we have seen, there are alternatives. Bergsonian philosophy abstains from thinking the universe in terms of the collection of all particles (to paraphrase Jan Bor[20]), nor time in terms of the linearity of the first dimension.

In spite of his avant-garde approach to physics, Einstein was in many ways a traditionalist. New as this synthesis of space and time may have been, Einstein's special theory of relativity fails to completely abandon seventeenth century mechanics because of its tenacious attachment to clock-time. It rather serves to save this framework. His ideas lean heavily upon Newton's mechanics, and are often led by theoretical assumptions and mathematics rather than experiment – here as well he differed radically from Niels Bohr. Einstein opposed quantum theory since he saw it as incomplete and inconsistent. He retained a worldview in which the world appears as a collection of separable and perceivable objects with definitive locations (Eastman 2004). This is the ultimate reason for him to oppose quantum theory, specifically in how it was articulated in the Copenhagen interpretation of quantum mechanics.

In the Copenhagen interpretation devised between 1925 and 1927 by Bohr and Heisenberg, the physical world is taken not to have any definitive properties until they are measured. The consequences of this curious position is that, ultimately, reality is dependent on the observer. Furthermore, it takes an ontological position with regard to reality as consisting of probabilities rather than actualities. This was necessary to be able to account for the double nature of light as both wave and particle (photons). Whilst classical physics entails the view that particles are clearly distinguishable from waves, this is not the case for quantum theory. Here, there is an ambiguity between waves and particles. This wave-particle-dualism is a logical paradox, that is empirically confirmed. The wave-particle descriptions of some states of nature are just that – descriptions. The underlying reality that there are phenomena which comply with either description is dependent of the observer.

---

[20] Personal dialogue with Jan Bor, during a car trip to the Auvergne, on the criticism of the conception of the world as the collection of all things.

3.3 Bohr, de Broglie and the Unfreezing of Time

After reading the previous paragraphs one might be tempted to accuse Einstein of philosophical naivety as it was for many in the past to accuse Bergson of scientific naivety. But to do so would be a mistake: Einstein was well aware of the philosophical arguments that were relevant to his work. He did however not take the effort to truly engage in Bergson's arguments for quite some years. His motifs may have been personal, they may also have consisted of a mere lack of time (as witnessed by his letter from the ship to Japan on *Duration and Simultaneity*). All we can do now is to try and reconcile both readings of the problem of time. This paragraph will attempt this and discuss how this may also be relevant for the gap between two dominant paradigms in twentieth century physics that I introduced at the end of the previous section: Einsteinian relativity and Bohrian quantum mechanics.

The basic viewpoint of Einstein and the basic viewpoint of Bohr informed two theoretical paradigms that dominated twentieth century physics: the general theory of relativity and quantum theory. They differ fundamentally in one respect: where relativity theory remains deterministic, quantum theory is indeterministic.[21] Quantum physics describes the behaviour of small particles such as atoms and electrons, while relativity theory accurately explains cosmic forces, but in some cases, notably with regard to gravity, the two theories seem incompatible. Whilst Einstein implicitly remained Cartesian at heart, Bohr went beyond.

Both paradigms are troubled by their conception of time. And this problem endures. As the physicist Lee Smolin states:

> "I believe there is something basic we are all missing, some wrong assumption we are all making. […]. My guess is that it involves two things: the foundations of quantum mechanics and the nature of time. […]. More and more, I have the feeling that quantum theory and general relativity are both deeply wrong about the nature of time. It is not enough to combine them. There is a deeper problem, perhaps going back to the origin of physics." (Smolin 2006, pp. 256–257)

The problem separating the two dominant paradigms in physics is best illustrated by the double-slit experiment. In this experiment, a light source (e.g. a laser beam) is directed at a plate with two parallel slits. The light passing through the slits is projected onto a screen behind the plate. Light is considered to be a wave, and this is confirmed by the wave interference pattern that appear on the screen (in the form of light and dark bands). If light would consist of particles, one would merely expect the projection of two slits. However, if one studies the way in which the light lands on the screen, by allowing one photon at the time to pass through the slit, light appears to consist of discrete points, as one would expect if light consists of particles. Furthermore, whilst waves would pass through both slits, particles would only pass through either the one or the other, and oddly, the latter is confirmed in the experiment.

---

[21] Although this indeterminism is often taken to apply merely to phenomena (the measurement), not necessarily to what actually is (what is supposed to be measured).

The double-slit experiment shows that light can display the properties of waves, in their classical definition, as well as particles, although having both properties is logically inconsistent. If one insists on identifying light in terms of particles with specific and unique locations at one moment in time, then one can only imagine such particles, in shooting them at random at a wall with two slits, as going through one slit or another. The experiment demonstrates, however, that quantum phenomena are fundamentally probabilistic in nature. Although in relativity theory the nature of matter and particles is not fixed either, the relation between matter and energy is. The findings of the double slit experiment thus seem to be in direct contradiction with the basic deterministic principles that characterise Einstein's theoretical framework. They also show two very different attitudes towards physics research. The Copenhagen interpretation of quantum mechanics takes the position that photons sometimes seem like particles and sometimes like waves since they are products of our experiments and do not have independent reality. The Copenhagen interpretation of quantum mechanics takes the position that photons sometimes seem to behave like particles and sometimes like waves because they are artificial products of our experiments and do not exist as discrete, independent entities in reality. Bohr was aware of the problem but refused to give theoretical consistency priority over empirical results. Einstein never went along with this view. He never accepted the relativism implied.

It has been suggested that one might consider atoms as quantum wave-particle entities, but this suggestion does not really solve the conceptual issue. Quantum theory retains an ambiguity, wavering between waves and particles, a paradoxical wave-particle-dualism that is oddly enough confirmed by experiments. After decades of experiments and debates, the puzzle still has not been resolved satisfactorily. This is a reason for supporters of the Einsteinian framework to remain critical of quantum theory. Einsteinian relativity might have changed the idea that time and space are absolute, it does not embrace the idea that reality itself might not be deterministic. The gap between relativity theory and quantum theory hinges not only on one's conception of matter and particles, but also of the basic nature of change and causality, and thus time.

Erwin Schrödinger conceived of a problem with regard to making an exception of the quantum level: apparently phenomena on that level can behave both as waves and as particles. This apparent dualistic behaviour at the quantum level, led Heisenberg to the definition of the uncertainty principle, defined in 1927 by the German physicist Werner Heisenberg. This principle states that there are limitations to our ability to simultaneously ascertain the position and the momentum of a particle or to simultaneously ascertain the energy and position in time of a particle. In other words, the more precisely we know the position of a particle, the less precisely we are able to ascertain its momentum and vice versa.

Popularised conceptions of the principle of uncertainty often regard it as going against the determinist worldview where it is retained in the Einsteinian universe. This interpretation of the principle is generally regarded as flawed. The original conception of this principle did however say something about the limitations of human knowledge: the uncertainty principle relates to knowledge rather than the state of the universe itself: it refers to what a human (or other conceptualising mind) can know. It does not speak of the system as such as being undetermined. Still, with regard to the debate over the Copenhagen interpretation of quantum physics, fundamental uncertainty remains an issue.

In the debate between Einsteinian determinism and Bohrian indeterminism Erwin Schrödinger took an interesting position. Since he abhorred the indeterminist consequences of the Copenhagen interpretation of quantum mechanics. In the Copenhagen interpretation subatomic particles (such as photons or radioactive particles) existed as a combination of multiple mutually exclusive but coexisting states, with different possible outcomes. Therefore he defined the following thought experiment:

> "One can even set up quite ridiculous cases. A cat is penned up in a steel chamber, along with the following device (which must be secured against direct interference by the cat): in a Geiger counter, there is a tiny bit of radioactive substance, so small, that perhaps in the course of the hour one of the atoms decays, but also, with equal probability, perhaps none; if it happens, the counter tube discharges and through a relay releases a hammer that shatters a small flask of hydrocyanic acid. If one has left this entire system to itself for an hour, one would say that the cat still lives if meanwhile no atom has decayed. The first atomic decay would have poisoned it. The psi-function of the entire system would express this by having in it the living and dead cat (pardon the expression) mixed or smeared out in equal parts." (here quoted from Greenstein and Zajonc 2006).

Schrödinger's experiment illustrated how, if quantum mechanics was correct, microphysical events (the presence and absence of a particle) would have macrophysical consequences (the cat that is both dead and alive).

In light of the problematic nature of quantum mechanics described above, some physicists did return to Bergson. Louis de Broglie, who approached the issue by considering every object in the universe as a wave, later perceived of a parallel between this uncertainty principle and the ideas of Bergson (Bor 1990, pp. 82). Remarkably, already in *Matter and Memory* (Matière et Memoire (1988 [1896])), Bergson argued that it is impossible for us to understand change and movement if we try to fix the exact location of something. Although critical of some of the views developed by Bergson in *Duration and Simultaneity*, de Broglie did agree with Bergson's non-distinction between world and mind. Bergson's views on the mutually exclusivity of observing change and location at the same time is quite consistent with the quantum-mechanical assertion that knowing both location and velocity is impossible on the subatomic microscale and merely an illusion on the macroscopic level.

The de Broglie-Bohm theory (Bohm 1952) provides us with an alternative view. First defined by Louis de Broglie, then abandoned, it was later picked up again by his younger colleague David Bohm (1917–1992). Bohm was a physicist interested in how time is conceived in mathematical approaches to physics (Cosgrove 2012). In elaborating de Broglie's intuition, he tried to develop an alternative to the Copenhagen interpretation of quantum theory (Gao 2003). De Broglie had struggled over the choice between an objectivist and a subjectivist interpretation of quantum mechanics. In the end he explained particles on the basis of their wave function, but this resulted in the need to speak about degrees of likelihood of particles being in a particular location (Skribna 2001). Therefore, in his view, 'particles' exist everywhere, to varying degrees.[22] This would have pan-psychic consequences, to which I will get back later.[23]

Several later physicists showed an interest in Bergson's process-based view of time. Several attempts were made to reconcile Bergsonian views of time with Einsteinian physics. One interesting example is *Physics and the Ultimate Significance of Time*, an edited volume by David R. Griffin (1985), based on a conference of the same name that involved both physicists and philosophers. The conference papers on which this volume are based, touch upon some of the most important problems encountered in both contemporary and past deliberations, and many synergies and frictions between the exact sciences and the humanities are discussed.

Griffin's introduction to the volume sketches out three different conceptions of time. First, there is the convention that time as we experience it does not exist. Here, the view of classical physics is most prominent: experienced time is merely an 'epiphenomenon', the result of how our psyche works.[24] Present, past and future are predetermined; the concept of time only holds relevance for how we speak about events. Secondly there is the convention that although time is experienced, this is only the case for those entities in reality that can be considered as sentient beings (plants, animals, bacteria). In this view, time exists as a subjective category. Thirdly, there is the convention that time is the basis of everything, but in terms of processes. Many modern physicists are intuitively drawn to the first account, and a neuroscientist might be drawn to the second conception. Bergsonian time lies in between the second and third conception of time: time – as duration – exists, it is very real indeed and not merely a side-effect of sensory perception. It is neither an epiphenomenon of how our brain is wired or how we interpret the world, nor is time part of a fixed relation with either mass/gravitation or velocity. This third position has received increased attention.

In a philosophical postscriptum to his famous essay *What Is Life* (1944), Erwin Schrödinger states that quantum indeterminism should not be regarded as in some way providing a physical basis for free will, stating: "I wish to emphasize that in my opinion, and contrary to the opinion upheld in some quarters, quantum indeterminacy plays no biologically relevant role in them, except perhaps by enhancing their

---

[22] The de Broglie-Bohm theory, at face value, expresses a reconciliation between quantum mechanics and classical causal deterministic physics, thus forming an alternative to the Copenhagen interpretation's indeterminist implications. Later in his career, Bohm however stated that he preferred 'ontological' rather than 'causal interpretation' of quantum mechanics, since 'causal' resounded too much with determinism. In collaboration with Jean-Pierre Vigier and Basil Hiley, he elaborated this point and in *The Undivided Universe*, published a year after his death. Here, they developed further arguments for this non-determinist view (Bohm and Hiley 1993).

[23] Panpsychic that consciousness is everywhere.

[24] As an epiphenomenon, time would only be a mental projection that is derived from how the physical world appears to our mind. Time is a property that emerges when there are changes – without change nothing happens and time disappears.

purely accidental character in such events as meiosis, natural and X-ray induced mutation and so on -and this is in any case obvious and well recognized". But following de Broglie David Bohm would develop a different point of view. He picked up on the intuitions of de Broglie, comparing quantum processes with human consciousness. Louis de Broglie originally conceived of such an analogy, but eventually abandoned it. Bohm however took up this old position, stating: "[The atom] can best be regarded as a poorly defined cloud, dependent for its particular form on the whole environment, including the observing instrument. Thus, one can no longer maintain the division between the observer and observed. Rather, both observer and observed are merging and interpenetrating aspects of one whole reality, which is indivisible and unanalysable." (Bohm 1980, p. 12). Ultimately, Bohm would take the step to an assertion of panpsychism, stating "the implicate order applies both to matter and to consciousness" (Bohm 1980 p. 196, also cited in Skribna (2001)).

We may either opt for a subjectivist interpretation or for an objectivist interpretation of quantum mechanical indeterminacy. The Copenhagen interpretation of quantum mechanics implies the latter: reality is truly indeterminate in nature, and the duality is not merely a problem of our subjective experience of reality. Scientists who hesitate to choose between the subjectivist and the objectivist interpretation[25] also struggle with another problem: if reality is indeterminate, impying that the nature of events in reality depends to some extent on the observer, then the universe is panpsychic and comes into existence to the extent that there are sentient beings (such as ourselves) who are able perceive it. The universe can thus only come into being insofar as it is perceived. In other words, the world comes into existence as a dialectical interaction between subject and object. Although the idea that consciousness is able to bridge the Cartesian divide between subject and object seems attractive (Kreps 2015), this consequence of quantum mechanics explains why many physicists conceive of de Broglie and Bohm's views on the Copenhagen interpretation as excessively esoteric. Here it becomes clear that the implications of the erosion of the mechanistic time stretch much further than a mere disciplinary irreconcilability of theoretical frameworks.[26]

In physics, the struggle over time persists: in 2006, Lee Smolin suggested that quantum mechanics and general relativity cannot be reconciled if theoretical physics does not embrace an alternative concept of time to the static concept it uses up till now (Smolin 2006; Cosgrove 2012). Similar to Bergson, Smolin also criticises the persisting early modern tendency to think time in terms of space – relating this flaw to Descartes' and Galilei's invention of graphs with one axis representing time and

---

[25] such as the quantum physicist Louis de Broglie (1892–1987).

[26] Teilhard de Chardin, whose work I will discuss more elaborately in Chap. 5, wrote: "[S]ince the atom is naturally co-extensive with the whole of the space in which it is situated […] we are bound to admit that this immensity represents the sphere of action common to all atoms. The volume of each of them is the volume of the universe." (Teilhard de Chardin 1959, p. 45). As Skribna (2001, p. 284) affirms, this point is critical for Teilhard de Chardin's view of the deep unity of the cosmos. At the same time, one has to take into account the probability of an atom being at a particular position and it is only the notional probability that it could be anywhere at any particular time.

Image 3.1   Jackson Pollock, *Convergence* (1952). Oil on canvas: 237 cm × 390 cm

the other representing space. Neither conceived of time as geometrical in nature, but gradually this conception became more acceptable, becoming a kind of gospel through the concept of spacetime as it was drafted by Minkowski, in whose interpretation of relativity, time becomes a geometrical dimension. In Smolin's words, physics has to 'unfreeze time' if quantum mechanics and general relativity are ever to be reconciled (Smolin 2006; Cosgrove 2012). Physics remains overly 'akinetoptic': motion blind. The field all too often demonstrates an inability to register movement as movement since it retains a strong tendency to perceive of duration and flux as a series of separable slides. Thus, physics needs to liberate itself from its petrified views by becoming more open to seeing nature in terms of processes. This would however imply a complete transformation of both relativity and quantum thought.

Research in physics has, until now not been able to convincingly discredit the basics of Einstein's theory. But quantum theory and the theory of general relativity are still conceptually incompatible, especially if one chooses to embrace empiricism over logical consistency, as the Copenhagen interpretation of quantum theory did. Efforts to unite the two, such as string theory and loop quantum gravity, do appear to be promising alternatives to bridge the gap. In terms of both the presupposition of a separation between observer and observed – subject and object – and the problematic nature of a probabilistic view on reality (in terms of states that are in a continuous flux of becoming) it appears that Bergson's philosophical ideas might be relevant for developing such an alternative. Process thought might prove crucial to such radical transformation. Possibly, the disjunction between the two dominant traditions in physics, lies in their hidden metaphysical assumptions rather than the relation between physics theory and empirical settings. But whether the gap between the sciences and the humanities might also be resolved is a responsibility of both sides and requires in depth knowledge of the other side's stakes and positions. At the least, it should be noted that even Einstein did take a more positive viewpoint to Bergson's ideas. In a letter from 1953 to Michele Besso, he wrote: "You cannot get used to the idea that subjective time, with his "now", has no objective meaning. See Bergson!" (letter 197; 29 July 1953). Although the fragment in question does not articulate a deeper understanding of Bergson's treatment of the subject/object divide, Einstein did appear to have acknowledged that it is *objective* time ('*Temps*') that is an illusion; and indeed, experienced reality is an an immediate experience of change ('*durée*') and to understand change, one does not need clock time (see also Damour 2008, p. 12).

Herman Weyl, an old friend of Einstein who attended the 1922 debate later remarked something of elementary importance for the interpretation of the notion of vital impulse (élan vital). With vital impulse, remarked Weyl, Bergson described life as resisting the basic tendency of entropy. Whilst Boltzmann upheld an interpretation of the phenomenon of the direction of time as identical to the phenomenon of an increase in entropy (Klein et al. 1973), Bergson actually identifies time with life. Here, Weyl set up something that would later, through the works of Erwin Schrödinger, become known as negentropy. Schrödinger, in his aforementioned essay *What Is Life* (1944) speculated that the basic molecular structure of life might be compared to the (then recently discovered) structures of aperiodic crystals – crystals that demonstrate properties of a much more complex order than normal crystals studied by classical physics. In this reference, Schrödinger introduced a notion of life from a physicist's point of view, thus marking an important paradigm shift from physics to the life sciences. Bergson and Schrödinger seem to embrace similar notions of life: Schrödinger's notion of negentropy (negative entropy) as characteristic for living systems and Bergson's notion of élan vital seem quite interrelated. As said, Bergson's process-approach to life is different from the biological vitalist theses of the nineteenth century. Although asserting a difference between living and non-living nature, it does not posit them as different in kind, but rather as different in aggregation level. The next chapter will discuss the ruling paradigms on the topic of life in Bergson's day and age. It will then discuss the views of Bergson, in relation to a then emerging new paradigm in the life sciences. And finally I will demonstrate the relevance of these views for the present by extrapolating them to current and ongoing debates in synthetic biology.

Literature

Bergson, H. (1908 [1889]). *Essai sur les données immédiates de la conscience*. 6th ed., 1908, WS, IA. Doctoral dissertation.
Bergson, H. (1988 [1896]). *Matter and memory*. Translated bys N. M. Paul & W. S. Palmer. New York: Zone Books.
Bergson, H. (1911 [1907]). *Creative evolution*. New York: Henry Holt and Company.
Bergson, H. (1935 [1908]). *Bergson to William James*. May 9, 1908. Quoted from Perry, R. B., The Thought and Character of William James. 2 Vol. Boston: Little Brown and Company Boston, 1935. Vol. 2, pp. 623).
Bergson, H. (1965 [1922]). *Duration and simultaneity with reference to Einstein's theory*. Leon Jacobson; Herbert Dingle (transl.). Indianapolis: Bobbs-Merrill.

Bergson, H. (1999 [1922]). *Duration and simultaneity* (eds: Durie, R.), Manchester: Clinamen Press.

Bishop, M. J. (2003). *How to win the Nobel prize: An unexpected life in science*. Cambridge, MA: Harvard University Press.

Bohm, D. (1952). A suggested interpretation of the quantum theory in terms of hidden variables I. *Physical Review, 85*(2), 166–179.

Bohm, D. (1980). *Wholeness and the implicate order*. London: Routledge.

Bohm, D., & Hiley, B. J. (1993). *The undivided universe: An ontological interpretation of quantum theory*. London: Routledge.

Bor, J. (1990). *Bergson en de onmiddellijke ervaring*. Meppel: Boom.

Canales, J. (2005). Einstein, Bergson, and the experiment that failed: Intellectual cooperation at the league of nations. *MLN - Modern Language Notes, 120*(5), 1168–1191. Comparative Literature Issue (Dec., 2005).

Canales, J. (2016). Einstein's Bergson problem: Communication, consensus and good science. *Boston Studies in the Philosophy and History of Science, 285*, 58.

Čapek, M. (1971). Bergson and modern physics: A re-interpretation and re-evaluation. *Boston Studies in the Philosophy of Science, 7*. D. Reidel Publ. Comp.

Cosgrove, J. (2012). On the mathematical representation of Spacetime: A case study in historical-phenomenological Desedimentation. In B. Hopkins & J. Drummond (Eds.), *The new yearbook for phenomenology and phenomenological philosophy: Volume XI 2011* (pp. 154–186). Acumen Publishing.

Damour, P. (2008). What is missing from Minkowski's "Raum und Zeit" lecture. *Physics History*.

De Saint-Ours, A. (2008). *The rediscovery of time through its disappearance*. The Nature of Time Essay Contest.

Eastman, T. (2004). Duality without dualism. In T. Eastman & H. Keeton (Eds.), *Physics and whitehead: Quantum, process and experience*. Albany: State University of New York Press.

Gao, S. (2003). A possible quantum basis of panpsychism. *Neuro Quantology, 1*, 1.

Greenstein, G., & Zajonc, A. G. (2006). *The quantum challenge: Modern research on the foundations of quantum mechanics* (2nd ed.). Sudbury, MA: Jones and Bartlett.

Griffin, D. R. (1985). *Physics and the Ultimate Significance of Time: Bohm, Prigogine and Process Philosophy*. New York: State University of New York Press.

Hilbert, D. (1917). "Die Grundlagen der Physik (Zweite Mitteilung)", Nachrichten. Königliche Gesellschaft der Wissenschaften zu Göttingen. *Mathematische-Physikalische Klasse, 1917*, 53–76. [Hilber 1917 available online].

Holton, G. (1973). On the origins of the special theory of relativity. In *Thematic origins of scientific thought*.

James, W. (1975 [1907]). Pragmatism: A new name for some old ways of thinking. In F. Burkhardt et al. (Eds.), *The works of William James. Vol. 1*. Cambridge, MA: Harvard University Press.

Klein, M., Broda, E., & Flamn, L. (1973). In E. G. D. Cohen & W. Thirring (Eds.), *The Boltzmann equation, theory and application*. Wien-New York: Springer.

Kreps, D. (2015). *Bergson, complexity and creative emergence*. New York: Palgrave Macmillan.

Lane Craig, W. (2016). *Bergson Was Right about Relativity (well, partly)*. https://www.reasonable-faith.org/writings/scholarly-writings/divine-eternity/bergson-was-right-about-relativity-well-partly/. Last accessed on 17-10-2019.

Latour, B. (2011). Some experiments in art and politics. *E-flux Journal, 23*.

Lebowitz, J. L. (2008). From time-symmetric microscopic dynamics to time-assymetric macroscopic behaviour. An overview. In G. Gallatovtti, W. L. Reiter, & J. Yngavson (Eds.), *Boltzmann's legacy* (pp. 38–62). Zürich: European Mathematical Society.

Lettevall, R., Somsen, G., & Widmalm, S. (2012). *Neutrality in twentieth-century Europe: Intersections of science, culture, and politics after the first world war* (ed.). New York: Routledge.

Marij. (2014). On the origins and foundations of Laplacian determinism. *Studies in History and Philosophy of Science, 45*, 24–31. https://doi.org/10.1016/j.shpsa.2013.12.003.

Paty, M. (1986). Einstein and Spinoza. In M. Grene & D. Nails (Eds.), *Spinoza and the sciences. Boston studies in the philosophy of science* (Vol. 91). Dordrecht: Springer.

Petkov, V. (2005). Is there an alternative to the block universe view? *PhilSci Archive*. Retrieved on 20 Dec 2006.

Planck, M. (1897 [1903]). *Treatise on thermodynamics* (trans: Ogg, A.). London: Longmans Green.

Schrödinger, E. (1944). *What is life – The physical aspect of the living cell*. Cambridge: Cambridge University Press.

Skribna, D. (2001). *Participation, organisation, and mind: Toward a participatory worldview*. Doctoral thesis. University of Bath.

Smolin, L. (2006). *The trouble with physics: The rise of string theory, the fall of a science, and what comes next*. Houghton Mifflin.

van Dongen, H. (2014). *Bergson*. Amsterdam: Boom.

Weyl, H. (1918). Reine Infinitesimalgeometrie. *Mathematische Zeitschrift, 2*(3–4), 384–411. https://doi.org/10.1007/BF01199420. Reprinted in Weyl 1968a: 1–28.

Whitehead, A. N. (1960[1925]). *Science and the Modern World*. New York: Mentor/New American Library.

Chapter 4
Life and Time: Bergson and the Life Sciences

Life does not proceed by the association and addition of elements, but by dissociation and division. We must get beyond both points of view, both mechanism and finalism being, at bottom, only standpoints to which the human mind has been led by considering the work of man.

- Henri Bergson, Creative Evolution (1911 [1907])

In *Creative Evolution* (1911 [1907]) Henri Bergson discusses the distinction between the living, biotic, and the non-living, abiotic world. The study touched directly upon several of the key developments of his age in biology. Notably evolution theory played a crucial role in his philosophy. In the historical context of the turn of the century, the debate between religious views on the origin of life and evolution theory had far from quieted down. Evolution theory continued to be seen as controversial due to its negation of the story of the origin of species as it was laid down in Genesis. As such, the title *L'Evolution créatrice* hinted at a reconciliation of these opposing viewpoints.

Creative Evolution fleshes out Bergson's arguments against mechanistic interpretations of evolution theory. Bergson begins by taking a step backwards, taking the discussion back to basics. By discussing definitions of life as an emergent phenomenon, his work can be considered as a predecessor of complexity theory. It also addresses several issues in relation to the life sciences that continue to be relevant to current debates in this field. Notably his view on the origin of life forms an important potential contribution to current debates in the molecular life sciences.

For Bergson it was not sufficient to merely look at the material properties of living structures. A more comprehensive perspective to be able to account for the phenomena of life. At first glance, his views seem to place him dangerously close to vitalist biological theories of the nineteenth century. But on closer inspection they differ from biological vitalism on crucial points. Biological vitalists suggested that some kind of vital principle, or vital mystical spark makes life different from non-living structures. But Bergson's aim was not to explain the nature of life by invoking an unexplainable mystical factor. He instead aimed to show how change itself is a basic property of life. Not only life, everything in the universe is

© Springer Nature Switzerland AG 2021
L. Landeweerd, *Time, Life & Memory*, Library of Ethics and Applied Philosophy 38, https://doi.org/10.1007/978-3-030-56853-5_4

continuously changing, evolving and assuming different shapes. Everything there is diverging and converging in individuating movement. To phrase it in non-Bergsonian terminology, complex systems display autopoietic, self-creative, properties. As such, life indeed seems to display a specific tendency towards development and growth: a vital impulse that allows living entities to withstand and overcome the basic processes visible in of non-living matter such as gravitation, erosion, etc. Life thus seems to bear display an emerging property that cannot be found in the non-living world. To be able to explain what this distinction amounts to, it is best to turn to the study of the origin of life here on earth.

Material evidence of the beginning of life can be found on Akilia Island, off the coast of Greenland. The rocks of Akilia Island are amongst the oldest sedimentary rocks ever to have been found on earth. Dating back at least 3.85 billion years, these rocks were formed at a time just after a period in which large meteorites formed in the earlier stages of the solar system continuously impacted the earth's surface, and just after the surface of the earth had cooled down to form a stable crust. In spite of the harsh circumstances of this era these rocks already contain traces of life. The study of the origins of evolution, or abiogenetic biology, is the study of how life emerged. In the shape that we know life, on earth, it emerged from the chemistry of carbon, nitrogen, oxygen, phosphorous, sulphur and water.

From antiquity until after the Middle Ages scholars interested in the question of life would usually look at the works of Aristotle and others from his school. In these works, it was assumed that life could emerge spontaneously. But in the late seventeenth century, the theory of spontaneous generation was disproved by Francesco Redi and his findings were confirmed by Louis Pasteur. But these observations, whilst answering important empirical questions, did beg the questions where life came from if not spontaneously emerging, how it had started to evolve in the first place, and what set life apart from the non-living material universe. Furthermore, thermodynamics' second law later defined a universal principle of decay and life was a phenomenon that could not be easily accounted for if the insights of thermodynamics were taken as valid on both local and universal levels[1]: whilst thermodynamics informed the view that processes in the universe tend towards ever-increasing entropy and disorganisation, the phenomenon of life seemed to bear witness to an opposite process of increasingly complex forms of organisation. Essentially, life appeared to be a form of inversed entropy, thus adding to the mystery of its nature.

Leaving aside theories concerning intelligent design, some authors have speculated that life might have emerged in different places in the universe, and that life on earth was a consequence of interplanetary contamination (Wickramasinghe 2011). This so called 'panspermic' view is expressed amongst others by the geneticist Francis Crick, co-discoverer of the double helix. Crick, in his earlier works, even

---

[1] In discussions with an expert on thermodynamics, he defined a solution to this: the processes expressed in living beings were triggered by and are kept in motion through the energy of the sun. In this regard, the earth is not a closed system. As such, life on earth is merely a side-phenomenon of the heat generated by the sun. The total net calculus of entropy still carries along an increase in the entropy in the universe.

conjectured that the spread of life might have been intentional (Crick and Orgel 1973). Appealing as this xenogenic speculation on the origin of life on earth may be for the more imaginative amongst us, it still leaves open the question how life might then have emerged elsewhere in the universe.

Life scientists working from a systemic perspective often take a different, less mechanistic view on the phenomenon of life than molecular life scientists. Alexander Bogdanov (1873–1928), a Russian pioneer in systems thinking, and Karl Ludwig von Bertalanffy (1901–1972), an Austrian biologist, were two precursors of the systems approach. Bertalanffy explicitly distanced himself from an application of thermodynamics to open systems such as life on planet earth (Bertalanffy 1934). Focused on modelling systems through mathematics and computation, the field of systems biology aims for a more holistic approach, in which a synthesis of different disciplines is used to uncover the complex processes involved in living systems. The aim is not to reduce such systems to their constituent parts, but to study how certain properties of living systems (cells, tissues, organisms) emerge from the systemic whole. It does not focus on the aforementioned attempt (e.g. Evans 1951) to analyse the organisation of biology from the lower aggregation level of atoms, molecules and biomolecular complexes, through cells and tissues up to the higher level of organisms and ecosystems. It rather takes the apparent properties of the complexes of living systems as such as focus of observation and as a basis of explanation.

Another solution to the mystery of the origin of life is supported by many modern molecular life scientists. They insist that the problem of the origin of life is based on a mistaken assumption. They question whether life as such can objectively be regarded as distinct in nature from the inorganic world at all. In this view there is no categorical distinction between organic and inorganic matter, and we will be able to demonstrate this as soon as we are able to build life from non-living components ourselves (Church and Regis 2012): life, in its molecular structure, merely bears witness to a more complicated organisation of atoms, leading to more dynamic molecules and processes. Life is thus not fundamentally distinct from other, less complicated structures such as crystals or carbon molecules. This latter view is the result of specific progress made in the life sciences during the twentieth century: the marriage between neo-Darwinian evolution theory on the macro-scale and molecular chemistry and physics on the micro-scale. Evolution could now be reduced to its underlying mechanistic properties. As a result, many scholars discarded with the value of Bergson's work for the life sciences.

The marriage between evolution theory on the one hand and molecular chemistry and physics on the other appeared to graft the life sciences on chemistry and physics: the negentropic organisation of life could now be explained by and reduced to the basic insights of classical mechanics. As a result, many contemporary life scientists perceive of the processes of life and evolution as merely a more complicated form of chemistry. It also revives the ancient view of spontaneous generation of life: although the supposed 'mystery of life' is indeed a rarity in the immensity of a abiotic universe, there is at least the possibility of this rare occurrence of spontaneous generation. It must have occurred on earth, under conditions provided in the proverbial primordial soup. But it remains a phenomenon that can be reduced to basic molecular compounds. The properties of the systems involved already offer an explanation of the emergence of life: spontaneous variation of a complex and relatively instable molecule, combined with the selective processes of dynamic environments, further explains the diverse traits in living organisms. This approach fits with the mechanistic approach of modern science. It does not include, however, a convincing explanation of the self-organising properties of living matter or bio-molecules.

Molecular life scientists tend to follow the hierarchy of biological organisation mentioned above, whilst systems biologists aim to explain properties of complex biological systems through complexity theory. Complex systems theory shies away from teleological accounts of complex systems: there is no goal that predetermines the system to evolve towards it, although this may seem the case; but there is an emergence of new properties. Evolution theory is the quintessential example and most articulate approach to such a theory. Still, teleological accounts of evolutionary processes continue to sneak into the vocabularies used – while sometimes teleology (and theology) is explicitly added, intelligent design being one of the most stringent current examples. Biology still use the notion of function as being designed for purpose. In those cases, life is still treated as a process that follows some kind of pregiven path (even if only a 'strive to become sufficiently fit to survive').

The topic of the emergence of life continues to be a theoretical minefield. It is clear that contemporary (micro)biology still struggles with the issue. Research fields such as germline modification, CRISPR-Cas, synthetic biology etc. use an engineering approach to the analysis and synthesis of organic structures. Often, the engineering perspective frames living structures as if they were mechanical in nature. - Even though most bioengineers will admit that their approach to life cannot be equated with civic engineering practices, the engineering perspective does frame the debate on the nature of life.

The metaphor of engineering is used quite often in definitions and descriptions of these different fields. The ability to make something is increasingly equated with the ability to understand it. Here, Feynman's famous last dictum,[2] influenced a whole generation of life scientists: "What I cannot create, I do not understand". As van den Belt states: "In informational terms, Feynman's dictum boils down to Von Neumann's motto "If you can't compute it you don't understand it!". New life forms can be designed by writing 'programs' in the quaternary 'code' of the four DNA nucleotides. Hence, the work of synthetic biologists also resembles that of software designers." (van den Belt 2009): when we are able to recreate something, we understand it.[3]

---

[2] The physicist Richard Feynman wrote down the dictum with chalk on blackboard just before his death. Although he meant it to refer to the (re)creation of the argumentative steps towards a theoretical result, it has often been used to refer to the literal recreation of something in nature, to be able to understand it.

[3] Feynman is actually quite a bit older. Its historical roots can be traced to statements by Giambattisto Vico and Immanuel Kant (van den Belt 2009) in their focus on mathematical construction (Broeks and Zwart 2021).

Feynman's adage seems to be endorsed by many in the field of synthetic biology. Other life scientists involved in these fields reverse the consequences of this dictum. They feel that although the techniques used apparently work, they are still at a loss to explain why they work in the first place.[4] Thus, it remains unclear to what extent the synthetic biologist is an engineer or a bricoleur posing as an engineer. This latter position is more comparable to the practice of tinkering. Life as such cannot be engineered, but we can try to adapt a living system by tinkering with it here and there.

As outlined in Chap. 2, a key concept in Bergson's philosophy is the concept of vital impulse (*élan vital*). Vital impulse in Bergson's thought is directly interlinked with his criticism of both mechanistic and teleological accounts of life and evolution (he discusses the matter in Chap. 1 of *Creative Evolution* (1911[1907])). To be able to explain the meaning of Bergson's vital impulse for the philosophy of the life sciences, we need to take a few steps backwards, starting with the notion of causation as it was defined by Aristotle.

In Physics II-8, Aristotle explains that four causes determine the way in which things in the world change and evolve. These determine the material aspects of things (*causa materialis*), the formal aspects of things (*causa formalis*), the principles underlying things (*causa efficiens*) and the goal or final shape things strive for (*causa finalis*). Aristotle's final cause lies at the basis of the teleological worldview: the idea that things have a predetermined goal/cause that steers their becoming and their path. Teleology is a central aspect of Aristotelian philosophy. For Aristotle, the shape of things was underdetermined if one would only take into account the material, formal and efficient (principled) causes. In his analysis, the process of 'becoming' – growth, death, recurrence – necessitated a fourth cause: the idea that things have an intrinsic drive to reach a certain optimal shape. In Bergson's view, the focus should be on change itself, not as a derived property but as a first order reality. This necessitates a different account of the world. In other words, the focus is on the process of becoming, but without the Aristotelian teleological dimension.

It is tempting to equate Bergson's conception of vital impulse with Aristotle's final cause. But Bergson is adamant in his denial of the traditional interpretation of teleology, and the way in which final causality is embraced in this interpretation. In *Creative Evolution*, he criticises mechanistic thought on the basis of its negation of change, since it takes a determinist position, but he is critical of teleological accounts of life and evolution for a similar reason: teleology embraces the concept of a final cause, already present in previous states, and therefore equally deterministic. For it implies that all later states are already predetermined in former states. In Bergson's view this is a mistake made throughout modern science. The nature of life is, however, not fully explained by his 'reversed teleology' (see Sect. 2.1). There is a tendency in life to evolve into increasingly complex entities, which is why Bergson introduced a predecessor of a more modern notion of the 'complexification of life'.

---

[4] Personal communication with Prof. Vitor Martins Martins dos Santos (2017). Wageningen University and Research.

Different interpretations of evolution persist in classical biology, biochemistry, systems biology, synthetic biology, etc. These can be traced back to the history of the life sciences. They include the views on evolution of Jean-Baptiste de Lamarck (1744–1829), Herbert Spencer (1820–1903) and Charles Darwin (1809–1882), the views of Elie Metchnikoff (1845–1916) on the identity of the organism in relation to the immune system, Jöns Jacob Berzelius's (1779–1848) ideas on the origin of life, and much later Erwin Schrödinger's (1887–1961) views on life as negative entropy. They also include views that emerged with the advent of genetics and genomics from the 1930s onwards.[5] On the one hand, there is something like design at work: the genetic program. But this program evolves due to evolution, and not in a predetermined direction.

4.1 Mechanistic Thought and Evolution

Under the influence of the impressive mechanical dolls and automata produced in the seventeenth and eighteenth century, comparisons between organisms and machines became increasingly popular in the sciences: the anatomy of man and animal could be explained on the basis of the mechanical workings of machines and clocks; the function of the blood vessels explained by comparison to the rubber tubes that animated the lifelike dolls which mesmerised audiences in the seventeenth and eighteenth century. René Descartes still reserved some other realm for the human mind and soul, but one of the pre-Darwin adherents of materialism, Julien Onffray de La Mettrie (1709–1751), did away with such notions altogether, stating in his treatise *Man a Machine* (L'Homme Machine 2003[1748]) that "the human body is a machine which winds its own springs", while posing that the diverse states of our minds correlate with those of our bodies. From this perspective, the human being is a mechanism, and although we might seem to have a soul, and seem to demonstrate rationality and intentionality, these mental activities also remain dependent on physical (biochemical) causes.

Darwin's evolution theory itself takes a curiously hybrid position between such mechanistic views and process thinking. Whilst describing the way in which species evolve, these processes were still presupposed to ultimately rely on the laws of classical mechanics: the mutations giving rise to inheritable variability were still seen as dependent on pregiven mechanisms, complicated as these may be. Process thinking entails a shift of focus from determinism to emerging processes. The diversity of species should not be regarded as a diversity of fixed shapes, but a diversity that has emerged over time in processes of convergence, divergence and co-creation. For some the complexity involved is, in theory, reducible to mechanistic causally

---

[5] In early 1943, by invitation of the Dublin Institute for Advanced Studies at Trinity College, Schrödinger gave a view on biology in a series of lectures he later published under the title "What Is Life?: The Physical Aspect of the Living Cell". These were bundled in a small book with an enormous impact. It indirectly inspired the definition of the double helix (Schrödinger 1944).

determined processes, but for others the concept of emergence entails fundamental (that is, ontological) unpredictability. Here it is important to keep in mind that for the first, complexity is a pragmatic choice. For them, mapping the behavioural regularities that can be observed in complex systems is more practical than trying to reduce systemic behaviours to their causes in a linear, deterministic manner. Life is complicated, rather than genuinely complex. For those adhering to the second position, however, this is principally impossible. Complexity means that the future can not be predicted on the basis of knowledge concerning current states on a fundamental level, not due to our inability to take the position of Laplace's demon (see previous chapter).

The distinction between complexity and complicatedness (see a.o. Morin 1990; Cath 2018) also entails two different views of the concept of time. In fact, what we are dealing with here are two different, even incompatible, worldviews. These two worldviews are quite apparent in contemporary life sciences. On the one hand many biologists take an engineering perspective to both their technological innovative research and their scientific analysis of life. In this respect the mechanistic worldview still remains firmly established in biology and evolution is merely seen as the result of a series of pregiven and predetermined factors – a rearrangement of pre-existing elements.[6] But according to others the emergence of life cannot be reduced to mere mechanical principles (although still dependent of complicated and lengthy bio-chemical processes). Some scholars in systems biology embrace the idea that it is impossible to reduce complexity to complicatedness: complex systems are indeterminate in nature. In other words, although many scientists still adhere to the position that living systems are, more or less, like complicated clockwork mechanisms, which merely differ from their manmade counterparts in the material from which they are composed, others argue that the distinction is more fundamental. Mechanistic determinism and organic process thinking are two worldviews. They not only give rise to clashing scientific perspectives. They also inform opposite views of our relation to and place in nature, of how we should perceive the role of technology and innovation and how we should think about the future.

For Bergson, science was incapable of grasping the world in other than symbolic terms. As he wrote in his *An Introduction to Metaphysics*:

> "Now it is easy to see that the ordinary function of positive science is analysis. Positive science works, then, above all, with symbols. Even the most concrete of the natural sciences, those concerned with life, confine themselves to the visible form of living beings, their organs and anatomical elements. They make comparisons between these forms, they reduce the more complex to the more simple; in short, they study the workings of life in what is, so to speak, only its visual symbol or manifestation. If there exist any means of possessing a reality absolutely instead of knowing it relatively, of placing oneself within it instead of looking at it from outside points of view, of having the intuition instead of making the analysis: in short, of seizing it without any expression, translation, or symbolic representa-

---

[6] Bergson already criticised evolution theorists of his time for taking this position since to his mind they did not fundamentally rethink time in evolution theory, and thus still subsumed evolution under a Newtonian mechanistic principle (Bergson, *Creative Evolution*).

tion – metaphysics is that means. *Metaphysics, then, is the science which claims to dispense with symbols*" (Bergson's Italics, Bergson 1999 [1903], p. 24).

With positive science, Bergson refers to the positivist movement that flourished in the nineteenth century, also in philosophy. Important representatives were Henri de Saint-Simon (1760–1825), the aforecited Pierre-Simon Laplace (1749–1827) and Auguste Comte (1798–1857).[7] It was the aim of positivism to analyse events in reality in terms of their properties and relations. As such, it took all valid knowledge to be a posteriori, coming to us via sensory experience. Its aim was to replace metaphysics, and it is on this point that Bergson perceives a basic flaw in positivism: it mistakes its own symbolic representations for genuine rational knowledge.

The most relevant contribution Bergson made to the debates in biology of his time is probably outlined in *L'Évolution creatrice*. At a time when a fierce debate existed between creationists[8] and evolution theorists, a title like this could hardly go unnoticed. His main point, for this debate, consists of his application of the concept of *duration* to all processes of life, including the genesis of species. Biology was at that time still dominated by taxonomy and scientific practice consisted of collecting specimens and identifying new species. Eventually, however, these observations and reflections on taxonomy gave rise to other perspectives on the nature, other views on the variety of species and its origin.

Bergson criticised the Darwinian view on evolution (that had several predecessors) since he regarded it as mechanistic and artificial. His criticism focused on the role of chance and natural selection in Darwin's theory. His main point was that to understand the phenomenon of life, one cannot take an outsider's perspective and reduce the phenomenon to its basic properties: variation and selection. If one does, one only grasps those aspects of life that are not typical of life. To genuinely understand life, we must understand it from the inside, as a process in which we ourselves are always already involved (thinking about life, for instance, is itself already a phenomenon *of* life). As Paul-Antoine Miquel states: "[T]he hypothesis of natural selection is constructed in the reverse direction: we must understand how and why the action of nature is not intentional, like the action of a breeder of domestic species" (Miquel 2007): in *Creative Evolution* Bergson discusses life as an immanent force (1911 [1907], p. 149). As a consequence, life needs to be understood from within. Vital impulse is the aspect of life that makes it possible to overcome obstacle and constraints (such as entropy) and to attain higher levels of complexity than is possible in the abiotic realm. Likewise, mental existence is not a separate realm, but part of the process of life itself, and it does not restrict itself to the consciousness of humans. Rather, human consciousness itself is a specific manifestation of evolutionary emergence, which already contains traces of consciousness. When speaking

---

[7] The aim here is not to dispense with symbols, but rather to explain how the (symbolic) conventions within given fields of knowledge may lead to overstretched conclusions, and thus create openings for further theory and research.

[8] Creationism departs from the view that all species were, at one point in time, at the beginning of the history of the cosmos, created, and remained unchanged in the ensuing centuries.

about the obstacle of matter, Bergson refers to the restricted nature of non-living matter in terms of its restricted variance of divergence and convergence. Life thus cannot be regarded either as having a telos, a purpose, or s being merely a mechanistically determined phenomenon. Life is a process, giving rise to emergent phenomena and properties, and gradually increasing in complexity.

Image 4.1 Flower petal nests made by *Osmia avosetta*, a solitary bee species

Life is typified by a 'tendency'. It is not restricted to the present, nor directed by the past. It is, in Bergsonian terms, oriented towards a future which cannot be determined in advance. Living things demonstrate a non-thingly aspect in their existence: they unfold over time, rather than being confined to their present state. And whilst this might also be valid for the birth of stars or the behaviour of atoms, living structures are, if understood from the inside, not governed by the same overall principle. Life does not merely 'take place'. It is neither 'created' (be it by an omnipotent divine being or some other, perhaps alien, form of intentionality) nor haphazard (in terms of randomness). Life is neither created, nor predetermined, but crea*tive*. It therefore is not governed by the classic concept of the arrow of time. At the same time, life cannot be regarded as a separate metaphysical category. Life in Bergson's oeuvre, appears as a phenomenon that, although not fundamentally different from non-life, is different in its complex unfolding over time. Life is autopoietic: it creates itself. And it is on this point that Bergson's philosophical vitalism differs strongly from biological vitalism in its nineteenth century biological meaning.

4.2 The Vitalist Thesis

As said, before and during the nineteenth century, the bulk of biological research consisted in mapping the variety of species rather than in explaining its occurrence as such. Its focus was on taxonomic subdivisions. And although the subdivision of the different kingdoms of life (animals, plants, etc.) developed in a context of creationism was improved upon, the most important aspect of this renewed attention to the variety of species was ignored; namely that the evolution of species is a spontaneous, organic phenomenon. It does not follow some kind of aprioristic plan. For some evolution theorists, amongst whom pioneers such as Herbert Spencer (biologist and autodidact philosopher), this led to the idea that evolution was driven by predetermined principles. In Spencer's view, the principles of evolution followed the 'grace' of economic effort (Spencer 1854, p. 382): thus the chart horse was to be considered less successful in evolutionary terms than for example a racing horse. Spencer explained this aesthetic view on evolution on the basis of the idea that the most economic motion would also be the most graceful, and thus the most evolutionary successful one. In his earlier works, Bergson was strongly influenced by Spencer. But Spencer's interpretation of evolution implied that it had a plan or purpose. It was moving towards a final cause. In 'Time and Free Will' (p. 11), Bergson therefore turned away from Spencer's views, as James also did, and focused on an alternative account for the concepts of time and evolution. He critiques Spencer for "reconstructing evolution with fragments of the evolved": in other words, Spencer reduces evolutionary processes to yet again mechanistic relations.

The taxonomic drafting of different classifications was basically compatible with Biblical views on life. Nineteenth century science museums are filled to the brim with collections of samples of elements in the periodic table, specimens collected by crystallographers, biologists and ethnographers.[9] The sciences themselves had also been classified and throughout the nineteenth century disciplinary borders remained firmly established. As a result, physics was defined as a field that stood apart from chemistry, and chemistry as standing apart from biology. To maintain these borders, and safeguard a special place for life within the wider scope of creation, many biologists adhered to vitalism. At the time, there was still some space for criticising Darwin's theory of evolution from a rational-empirical point of view: in spite of the apparent rationale in his theory, the timescale of evolution was too immense to allow for experimental verification, whilst the molecular basis for spontaneous variation had yet to be uncovered. The original vitalist thesis proposed by the chemist Jöns Jacob Berzelius held that organic life could not be synthesised from organic components. They therefore adhered to the idea that some regulative force must exist, and exist exclusively in life forms.

One problem evolution theorists were confronted with was how to explain the emergence of life as such: how could organic life forms have emerged from inorganic compounds? Although vitalism is largely discredited by now, the debate on the origin of life around the early 1900s was at its peak: either life was no different from nonlife, and should thus be reducible to its various chemical components and their physically determined behaviour; or life was of an entirely different category due to some kind of vital spark or energy, and thus it escaped such physico-chemical determinism. The backdrop of this debate was formed by the implicit materialist

---

[9] At the time also with 'specimens' that remain a horrific witness to humanity's talent for cruelty in the name of science and truth.

consequences of a non-vitalist position; after all, if organic materials can be composed from merely inorganic components, they can be reduced to these as well. Life, in essence, would be nothing more than a more complicated constellation of material components. It would be determined by the forces that equally govern the inorganic world. Human nature (including the problem of free will) constituted the centre of the vitalist debate. But the idea of some kind of mystical life force could not hold up to scientific scrutiny.

The relevance of Bergson's *Creative Evolution* for this debate lies in his emphasis on the interrelation between vital impulse (in terms of duration) and the phenomenon of evolution. The emergence of a wide variety and ongoing variation of species cannot be regarded outside of time. Darwinian theory presupposes that the process of evolution has enormous expanses of time at tis disposal. Evolution requires deep time rather than clock time. Bergson:

> "The present state of an unorganized body depends exclusively on what happened at the previous instant; and likewise the position of the material points of a system defined and isolated by science is determined by the position of these same points at the moment immediately before. In other words, the laws that govern unorganized matter are expressible, in principle, by differential equation in which time (in the sense in which the mathematician takes this word) would play the role of independent variable. Is it so with the laws of life? Does the state of a living body find its complete explanation in the state immediately before? Yes, if it is agreed *a priori* to liken the living body to other bodies, and to identify it, for the sake of the argument, with the artificial systems on which the chemist, physicist, and astronomer operate. But in astronomy, physics, and chemistry the proposition has a perfectly definite meaning: it signifies that certain aspects of the present, important for science, are calculable as functions of the immediate past." (Bergson 1911 [1907], p. 19).

Here, Bergson – in spite of himself[10] – embraces a Kantian strategy for the sake of the argument. How far can we stretch the mechanistic logic of positivist science? He regards mechanistic determinism as an a priori condition for specific approaches in the sciences, which may be functional to some extent, but also excludes specific aspects of certain phenomena. It is wholly inadequate, however, when it comes to explaining life. Indeed, as Bergson continues:

> "Nothing of the sort in the domain of life. Here calculation touches, at most, certain phenomena of organic *destruction*. Organic *creation*, on the contrary, the evolutionary phenomena which properly constitute life, we cannot in any way subject to a mathematical treatment. It will be said that this impotence is due only to our ignorance. But it may equally well express the case that the present moment of a living body does not find its explanation in the moment immediately before, that all the past of the organism must be added to that moment, its heredity – in fact, the whole of a very long history. In the second of these two hypotheses, not in the first, is really expressed the present state of the biological sciences,

---

[10] He objected to Immanuel Kant's epistemological conception of time (*Critic of Pure Reason* (*Kritik der reinen Vernunft* (1781/1787)) since Kant, as mentioned previously, seemed to interpret time as a spatial dimension. There is reason to draw this assertion into question, since for Kant, time is very much constituted by our ability to recognize a thing as the same thing over time. This is amongst others captured in Kant's 'modality of recognition': our ability to recognize something as something relates time to experiential memory rather than to space: Kant thus reconstructs time as an aspect of memory, not of space, a strategy quite similar to Bergson's in 'Matter and Memory'.

as well as their direction. As for the idea that the living body might be treated by some superhuman calculator in the same mathematical way as our solar system, this has gradually arisen from a metaphysic which has taken a more precise form since the physical discoveries of Galileo, but which, as we shall show, was always the natural metaphysic of the human mind. Its apparent clearness, our impatient desire to find it true, the enthusiasm with which so many excellent minds accept it without proof – all the seductions, in short, any attempt to distinguish between an artificial and a natural system, between the dead and the living, runs counter to this tendency at once." (Bergson 1911 [1907], p. 20, my italics).

It was for the reason of such a process-based approach that Bergson preferred the account of evolution provided by Lamarck over the account provided by Spencer or, for that matter, over some aspects of Darwin's account of evolution.

Darwin's theory of evolution did not allow for plasticity on the level of the individual. Whilst Darwin's view on evolution remained dominant throughout the twentieth century, recent research in microbiology confirmed that some plasticity of the genome during an individual lifespan does occur, and can indeed be transferred to offspring. However marginal these effects may be, in the grander scope of the timescale of evolution, such intra-individual variation does contribute to the diversification of life. And whilst the elegance of Darwin's approach to evolution relies on the idea that variation occurring at the level of the species rather than at the level of individual phenotypic expression, is sufficient for evolution to occur, other formative principles are not excluded per se.

In Bergson's ideas on evolutionary processes, the concept of vital impulse can philosophically speaking be compared to the concept of 'will to power' (*Wille zur Macht*) as Friedrich Nietzsche elaborated it on the basis of Arthur Schopenhauer's 'will to life' (*Wille zum Leben*): vital impulse also serves to explain the way in which life apparently strives to continue. Still, Bergson tried to avoid the idea that vital impulse alluded to a general cosmic inner force manifesting itself in the manifold. Vital impulse, as creative complexification over time, accounts for the differentiation of life forms. Individual plants are more or less stationary in principle, whilst animal life, lacking the function of photosynthesis, is mobile. Human life is again different, since it demonstrates intelligence next to instinct, creating language and tools. As Bergson's successor Édouard Le Roy (1870–1954) would say, human intelligence lies at the basis of a sphere of knowledge surrounding the biosphere, a so called 'noosphere'. In all these aggregation levels, different life forms become manifest in a process of individuation (see Sect. 4.3), but all in terms of divergence and differentiation on the one hand and convergence and unification on the other. In short, all these manifestations, including human consciousness, are part of the same creative process.

Living systems converge and diverge. This phenomenon exists on different aggregation levels: it is imperceivable but still slightly manifest at the inorganic level, but more so at the organic level. It is manifest in life, but also at the level of knowledge and technological development. Presently, the mechanical not only transforms the world around us, it also affects human nature: the mechanic, the artificial, is merging with the organic, the natural on a deep level. Technology is thus an intrinsic part of nature; evolution is becoming manifest in living technological systems: developments which are perfectly comprehensible from a Bergsonian point of view.

The vitalist position implicitly embraces a teleological view. If translated to contemporary life sciences research, a neo-vitalist position would entail that the 'programmes' contained in DNA serve as final cause of the ultimate phenotype. DNA is the 'concept' or eidos, which realises itself in the organism. This is why Aristotle has been credited with anticipating genomics (Zwart 2018). Most contemporary life scientists usually do not perceive of a life force other than the programme contained in DNA. For them, the DNA molecule is merely regarded as the programme informing the phenotypic shape of organisms. Therefore, one might wonder whether Bergson's vital impulse is superfluous, now that DNA proves to be the driver of the processes of life. But the history leading up to the conception of DNA reveals a much richer account. The difference between non-living and living, between entropy and negative entropy, is the DNA molecule, which changes in an adaptive manner in the course of evolution. This becomes clear when we turn back to Schrödinger's aforecited essay.

As mentioned, Schrödinger defined the molecular essence of life as an 'aperiodic crystal'. Crystals might be considered as paradigm examples of ordered structures in the material universe. Such order was once regarded as synonymous with whether the arrangements of the atoms involved, repeated itself sufficiently regularly throughout a crystal structure. These are the so called periodical crystals. Aperiodical crystals, defined after the discovery of quasi-crystals, are of a different, much richer and more complex organisational order. In Erwin Schrödinger's words: "The difference in structure is of the same kind as that between an ordinary wallpaper in which the same pattern is repeated again and again in regular periodicity and a masterpiece of embroidery, say a Raphael tapestry, which shows no dull repetition, but an elaborate, coherent, meaningful design traced by the great master." (Schrödinger 1944). Schrödinger applied this idea of the "aperiodic crystal" to the basic fabric of life. Life might be understood as a crystal-like ordered structure, containing genetic information in its chemical configuration. But this chemical configuration is infinitely more complex that the structure of periodic crystals. On this basis and on the basis of the experimental work of Rosalind Franklin, Watson and Crick arrived at the idea of a double helix (the molecular basis of living systems). Here, the position that Bergson adheres to in *Creative Evolution*, that crystals and organisms are distinct (1911 [1907] p. 12), is challenged, and remains problematically outdated: whilst crystals do not diverge in function, aperiodic crystals do, thereby representing a bridge between abiotic and biotic nature. There are similarities between the original fabrics from which life emerged and aperiodic crystals. But research into quasi-crystals and aperiodic crystals did not spur the scientific speculation that would lead to the hypothesis of the double helix structure of DNA until after Bergson's death.

In an epilogue to 'What is Life?' entitled 'On Determinism and Free Will', Schrödinger addresses another emerging property of living entities, notably in humans, namely consciousness. In reference to the Upanishads, Schrödinger ends with a note on the consequences of this conception of the essence life being an aperiodic crystal for the philosophical debate on free will and determinism. His view:

> "I believe every unbiased biologist would, if there were not the well-known, unpleasant feeling about 'declaring oneself to be a pure mechanism'. For it is deemed to contradict Free Will as warranted by direct introspection. But immediate experiences in themselves, however various and disparate they be, are logically incapable of contradicting each other. So let us see whether we cannot draw the correct, non-contradictory conclusion from the following two premises:

(i) My body functions as a pure mechanism according to the Laws of Nature.
(ii) Yet I know, by incontrovertible direct experience, that I am directing its motions, of which I foresee the effects, that may be fateful and all-important, in which case I feel and take full responsibility for them.

> The only possible inference from these two facts is, I think, that I — I in the widest meaning of the word, that is to say, every conscious mind that has ever said or felt 'I' — am the person, if any, who controls the 'motion of the atoms' according to the Laws of Nature." (Schrödinger 1944).

In other words, as Schrödinger continues, "I am God", or, in the Upanishads: Atman = Brahman: the personal self equals the omnipresent, all-comprehending eternal self. To put it in the terminology of Teilhard de Chardin: when 'I think', 'I' participate in a collective emerging phenomenon called thinking, a noosphere (see paragraph 7.2). Schrödinger's postscriptum deserves elaboration, but it opens up a space of reasoning that is usually not visited in the exact sciences. The stress on the need for a reconciliation between our immediate experience of consciousness and our empirical assertions concerning the molecular basis of life lies directly in line with the endeavours some decades before of Bergson.

In the 1944 lecture series from which the citations above origin, Schrödinger also sketches out that life 'feeds on' negative entropy, or negentropy, as the physicist Leon Brillouin (1956) would later call it (Crow 1992; Gould 1995). The metabolic system, responsible for change and exchange goes against the tendencies defined by thermodynamics. They seem to form an exception to its laws: life seems to contradict the increase of chaos in the universe. In taking a reversed interpretation of thermodynamics (in terms of negentropy) as a basis for this aspect of life, Schrödinger thus introduces a notion of change that is quite similar to the notion of life as a manifestation of time, as drafted by Bergson.

On the basis of the above, the distinction between organics and inorganics, or between life and non-living structures is clearly gradual, not categoric. This point of view was also embraced by Bergson. His ideas thus went directly against the classical vitalist thesis of an external force working upon matter to create life. Rather, this gradual difference entails a concept of evolution from one organisational aggregation level to another, reflecting an increase of complexity. Similar to the distinction between periodic crystals and aperiodic crystals, the amount of possible combinations grows exponentially, as does the level of complexity involved.

If one regards life as some type of special substance with a specific structure, this creates the need for some type of vitalist assumption. But if viewed as a process, there is continuity between organic (carbon based) materials and other forms of organised matter and life. Everything is based on the protons/neutrons/electrons of the energetic spectrum and the ways in which energy is concentrated and deployed, but the number of combination increases dramatically, so that there is more and more room for manifest creativity.

Bergson's concept of 'vital impulse' clearly differs from how preceding vitalist theories perceived life. He did not embrace the concept of a mystical spark of life that distinguishes the living organic from the dead organic and the inorganic. His concept of a vital impulse derives from another source: it leans on a difference between process and thing. Rather than taking the material thing as the primary nature of reality, he takes process as the starting point. Bergson held the view that non-living matter did not yet develop individuality – not having diverged into specific functions – whilst life did. Had this not been the case, his philosophy of life might be termed panvitalist, since it elucidates the process nature of the inorganic, the organic, the technological and the reflective sphere. The concept of a vital impulse does not seek to smuggle a superfluous argument into the discussion of the emergence of life, but rather to describe the nature of emergence as such. It focuses on how changing systems evolve, considering their increasingly complex manifestation. Rather than focusing on the material substances of the inorganic versus the organic, Bergson provided an explanation on the basis of organisational aggregation levels. Evolutionary thinking was, scientifically speaking, an important step. But one of the most important aspects of this renewed view on the genesis of species is that the evolution of species is a spontaneous, creative phenomenon that does not follow some kind of aprioristic plan.

Bergson suggested an explanation that necessitates taking an ontological step backwards: rather than starting from the assumption that the universe consists of matter, particles, confined to specific locations at specific moments in time, he suggests the universe consists of processes. Time, in this regard, is manifest in the ever-changing and ever-evolving nature of the universe. Here it becomes clear that his opposition to a reduction of time to clock time in physics is deeply intertwined with his views on the nature of life. The next section will explore the background of this shift of perspective, as well as its significance for contemporary life sciences.

Vitalism slowly seeped out of modern biology. Specifically, the discovery of the function of chromosomes in the 1930s appeared to make a vitalist explanation on the origin of life obsolete. Evolution appeared to be a mechanical process, not that different from any other mechanical process outside of biology such as the growth of crystals. Mechanistic metaphors thus guide the efforts of many contemporary life scientists, many of whom not only discard the notion that the chemistry of life is fundamentally different from inorganic chemistry, but also reduce the processes involved to, again, the mechanics of a clockwork.

4.3   The Alchemy of Life

A typical modern life sciences definition of life can be found in the works of Prof.
Raymond Spier, former editor of *Science and Engineering Ethics*:

> "One may consider a definition of a cell (in a biological context) to be a small entity, nor-
> mally between 0.1 to 100 μ in diameter, which provides an enclosed space. Life is more
> difficult to define as some individuals seek to associate immaterial properties (spirits, souls,
> life-forces) to living organisms (vitalism). If we do not do this, then a starting definition of
> life could take the form of – 'a living organism is an entity which 'seeks' (happens) to
> reproduce itself such that deviations in the reproductive process may be carried through into
> subsequent generations'. (In this expression the word 'seeks' has the meaning 'has the
> particular physical and chemical properties that, in the presence of the appropriate materi-
> als, spontaneously results in a reaction')." (Spier 2003).

Spier clearly shows a reluctance to use terms that might imply purpose, will or life
force. His definition attempts to demonstrate how life is only different from non-life
in gradual terms. Living organisms are merely more complicated organisational
structures, with no categoric difference from, say, a piece of granite. In this regard,
many modern life scientists still try to avoid any vocabulary that might hint at some
kind of special properties in life. 'Vitalism' is the key term here: in view of the fact
that vitalism is regarded as a childhood disease of the discipline of biology, it is
in need to a purgatory phase, ridding itself from the very word that defines it, to be
able to position itself between the more classical exact sciences. After all, these
themselves of esoteric and philosophical notions already in earlier stages of their
history (physics – necessitating an ignoring of a large portion of Newton's writings
(a.o. on astrology), chemistry – after its critical self-assessment in relation to
alchemy etc.).

As said earlier in this chapter as well as in the introducing chapters of this study,
vitalism in biology is known now mainly as the nineteenth century attempt to
reserve at least one sub-realm of creation for the soul. It is now regarded as pseudoscience. And, unfortunately, Bergson's ideas are still often placed in the same corner
as vitalism. But his philosophical vitalism is of a different nature. Bergson's philosophical vitalism (or 'philosophy of life') actually offers a solution to the problem
of biological vitalism.

The debate on what was life and what was not was still very much alive at the
beginning of the twentieth century and biological vitalism formed an important
theoretical frame in which spiritualism, theism and Darwinian evolution theory met.
Vitalism in biology was a reconciliation between science and a mystified deism.
And as a result, vitalism was taken less and less seriously in mainstream biology.
The term vitalism, still 'en vogue' in the early 1900s, linked the philosophy of
Bergson to nineteenth century vitalist biologists such as Berzelius. Berzelius's vitalism, however, suggests some type of mystical life force, a spark of life that distinguishes living from non-living systems. Similarly, vitalism considered vital impulse
as a mystical force acting on dead matter, thus separating the organic from the
inorganic.

Some aspects of vitalism persist in modern biology. The notion of a 'spark' of life is still used in contemporary discussions on the origin of life: one of the theses on the origin of life is that the polymers lying at the basis of the origin of life were created by the impact of lightening on the primordial soup. And xenogenetic explanations of the origin of life – that life on earth started with the impact of asteroids containing microbes from other parts of the solar system – are compatible with 'lightening stricking': it does not matter whether that 'spark' happened in the past on planet earth, or speculatively on another comparable planet nearby.

Apart from this 'defribilator'-view on the origin of life (or even at its cost) there is a persistent interpretation of evolutionary processes as having some kind of direction or goal. This persistence follows the notion that as it is ingrained in acorns to become an acorn tree, that plant life has 'progressed' when it invented flowers, that it was 'meant to be' that vertebrates have 5 digits (or, for some of them (e.g. horses), used to have them) and that apes of course evolved to move more and more upright and became bipedal.[11] This scientifically problematic teleological view on evolution implicitly embraces the idea that evolution is a process of progression. In general, older species are more primitive, less complex, than younger species. The teleological interpretation of evolution considers this to be proof of some kind of direction of evolution. This view is attractive to the human psyche, since it allows for a positioning of homo sapiens at the current end of the evolutionary chain: it undoes the 'narcissistic offence' implied in the replacement of creationism by evolution theory. But persistent as they are, teleological views on evolution demonstrate a deep misunderstanding of evolution theory since they introduce normative notions such as progress, primitiveness (rather than simplicity/complexity), etc. Even Charles Darwin himself did not succeed in keeping such normative notions at bay: his use of the notion of a 'struggle' for existence (inspired by the works of Thomas Malthus) does not seem to fit in with the emergence of a species like the sloth. At the same time, the originally neutral notion of a 'survival of the fittest', coined by Herbert Spencer, has come to accumulate connotations of physical strength, agility and superiority. Whilst the sloth is evolutionary quite successful, and 'fits' in its niche, it does not appear to be a very strong and agile creature: here one might say that the nature of each creature is a function of its habitat. And homo sapiens's bipedality is merely an expansion of that notion: from the invention of technology, using hands as a means to get somewhere became less important than using them to make 'stuff'.

Bergson was a vitalist philosopher, but of a different brand than vitalism as it was defined by Berzelius and others in nineteenth century biology. It might better be termed 'panvitalist'. The arrangements of matter, as Bergson states in In *Creative Evolution*, should not (yet) be conceived of as part of the order of life since only life individuates, only life generates individuality (Bergson 1911 [1907] p. 12). In *Creative Evolution* Bergson proposes an account of the principle of evolution as

---

[11] Something that is illustrated by the average, and often parodied, picture of the evolution of homo sapiens, from left to right, as a sequenced linear development from crawling dark-haired apes to increasingly tall, evermore upright (and, whilst disliking certain forms of overenthusiastic political correctness: indeed often also more blonde haired and lighter skinned) bipedal specimens.

motivated by a vital impulse, based on the chemistry of the nucleic acids whose properties include their self-replication given the availability of the materials used for this process. This impulse, in his view, also forms the basis of humanity's natural creative impulse. Life, from this view, is only allowed to manifest itself in sufficiently complex processes. Only then, convergence and divergence can occur at a level that is sufficiently creative for variation. The strive for self-organisation lies at the basis of the spontaneous morphogenesis of individuals and species.

Bergson's use of the concept of a vital impulse unfortunately led many to associate his ideas with those of biological vitalism. But in his work, vitalism is a specific perspective on the processes through which inorganic matter enters a higher state of organisational development, processes we refer to as evolution. The difference between biological vitalism and Bergson's understanding of creative evolution could not be bigger: biological vitalism demonstrates all the attributes of an Aristotelian teleological perspective. It sees the world as consisting of living entities and of things that are essentially inert and dead. Like billiard balls, planets and concrete bricks, these things cannot evolve out of themselves. They cannot move out of themselves, are subject to entropy, and remain fully unequipped for non-entropic change. From this position, the vitalist biologist proposes some type of external cause that acts upon these dead and inert things to create life. Bergson, however, positions himself explicitly against such teleological ponderings – as I discussed in the introductory chapters of this book. *Creative Evolution* entails a definitive denial of such a teleological thesis. He denies traditional Aristotelian and Leibnizian views on final causes – the idea that there is an external force that imposes its creativity on inorganic matter. The mistake implied in Aristotelian teleology is that it takes the discourse of mechanics (in terms of tools and cause-and-effect) and projects this anthropocentric perspective onto the natural world. What Bergson does say about the distinction between living organisms and dead inorganic matter is the following: "The only question is whether the natural systems which we call living beings must be assimilated to the artificial system that science cuts out within inert matter, or whether they must not rather be compared to that natural system which is the whole of the universe. That life is a kind of mechanism I cordially agree. But is it the mechanism of parts artificially isolated within the whole of the universe, or is it the mechanism of the real whole? The real whole might well be, we conceive, an indivisible continuity. The systems we cut out within it would, properly speaking, not then be parts at all; they would be partial views of the whole." (Bergson 1911 [1907], p. 46). Since it is vital impulse, understood as duration, that in the end puts life apart from dead, inorganic matter, duration is nature's tendency to converge and diverge into living entities.

For Bergson, the final cause should be positioned 'at the beginning': in other words, it is already implicit in the process-based way in which complex systems emerge and maintain their self-organisation. But it doesn't hold a predefined end-state. This self-organising principle is also manifest in human consciousness: experience, in its preconceptual sense, is always qualitative in nature. Teilhard de Chardin would elaborate this view in terms of 'evolution becoming aware of itself': the step from the emergence of living systems towards its manifestation in conscious and self-awareness in humans. This position is quite close to the position of David Bohm, whose views I discussed at the end of Chap. 3. Bohm stated: "The mechanical notion of an interactive universe is seen to be inadequate. It is in need of replacement by the notion of an objectively participative universe that includes our own participation as a special case" (Bohm 1980, p. 126).

Current developments in the life sciences appear to transcend the worldview that still governs the traditional objectivistic mentality. There is a growing tension between what scientists claim to endorse and what their research actually reveals, between the implicit metaphysics of science and the creative processes of emergence opened-up by current findings in the life sciences. The dividing line between living and dead complex structures is increasingly contested. But as a result, the concept of life itself is increasingly regarded as a term that belongs to a now defunct metaphysics. Synthetic biology, CRISPR-Cas, and other related paradigms and research domains are presented as a confirmation of the mechanistic worldview, with its insistence on material determinism. These developments rather confirm the creative dynamics of natural processes instead of disproving them. From the observation that the dividing line between living and non-living nature is blurry it does not follow at all that all life is 'mere' mechanics at work. One might even reverse the argument; it might also be defended that all systems in the universe show signs of life to some extent. The validity of these claims, however, cannot not be assessed in the postponed arrival of a renewed definition of life, whilstit is clear that traditional definitions of life are fundamentally challenged by what these newly emerging research fields reveal.

Life has mostly been defined in terms of metabolism, growth and reproduction. Although this seems fairly straightforward, it is not. Metabolism for example implies an interaction with an environment. And this again implies a differentiation between organism and environment. But this does not explain how life came to emerge from simpler, non-living organic molecules. On average, one can say that the more complex a molecular structure is, the less stable it will be. An exception is the molecular structure at the basis of life: here, molecules somehow appear to have developed a self-sustainable mechanism (on which I will introduce a lesser known account later in this section). It is at this point that the distinction between life and nonlife can be introduced.

Does this mean that the material determinism that we perceive in simpler, non-living physical processes does not also apply to the more complex processes of life that emerge at a higher aggregation level? Not yet: it only means that the 'white noise' that appears to a mere marginal extent in such non-living processes is much more manifest in living processes. This might be compared to what Carl Gustav Jung coined with the concept of individuation: the process through which, out of the amorphic and undifferentiated whole – think of the primordial soup, rather than the Jungian notion of the subconscious – something individual develops. The individual is then defined as 'not something else'. It has emerged autopoietically, as itself. Complexity is a precondition of life. Complex structures can only be regarded from a process-based perspective. Isolating one state of the system from time to analyse its intricacies still departs from the deterministic idea that the future of the universe is implied by, even already contained in the present. Complex systems thinking assumes the opposite, arguing that determinism only applies incrementally, according to the extent to which we consider a small part of reality outside its context and outside of time.

Bergson emphasised that we should regard the phenomenon of biological evolution from the perspective of duration. Evolution theory revealed that the difference between species emerged in a complex process of divergence and individuation. We all evolved from the same original organism. We only define a separation between species on the basis of the criterion of an inability to engage in fruitful reproduction. In this regard, evolution is a continuous process. In terms of the relation to the creationist debate, the ideas presented by Bergson tantalisingly propose a reconciliation between different perspectives. Evolution is a creative process, resulting in the emergence of new and increasingly complex species. Creation is better covered by a verb: it is constantly evolving, constantly revealing itself. This applies to all evolving systems, regardless of whether they belong to the organic world or to the world of technology.

Life, understood as time becoming manifest in changing matter, displays emergent properties. Contemporary life scientists are well aware of the debate on the nature of life and the distinction of the organic world and the inorganic world. Most embrace the view that there is no fundamental distinction between the realm studied by chemists and the realm studied by biologists. But epistemologically speaking, the divide used to be a significant or even insurmountable one. Chemistry studies elements and compounds (in their atomic, molecular, ionic etc. capacity). It studies the composition, structure, properties and behaviour of these elements and compounds. It specifically focuses on the changes these elements and compounds undergo if they react with other substances. The changes implied here are of a more basic nature than the changes studied in the life sciences. In the life sciences, self-organisation is a key element. Over the past decades, however, this clear division between chemistry and biology has all but evaporated. Many modern life scientists believe that the types of behaviour of living systems can ultimately be reduced to chemistry (and chemistry ultimately to physics). A paradigmatic shift has occurred, giving rise research areas such as biochemistry and the molecular life sciences.

On a more fundamental (metaphysical) level, however, these changes reflect a tension between two incompatible metaphysical paradigms, namely object metaphysics and process thinking. Lack of acknowledgement of this collision leads to misunderstandings and confusion in the life sciences with regard to the nature of evolution. Object metaphysics informs an interpretation of the world in terms of 'things'. Process thinking prioritises the temporal dimension, thinking in terms of processes rather than material things. Both paradigms can be found in the life sciences. Whilst the paradigm of process thinking embraces the view that life is a complex phenomenon that cannot be analytically reduced to its elementary components, the paradigm of object metaphysics presupposes that life is merely highly complicated. This implies that although the causal relations between the different atoms, molecules etc. that make up life might be difficult to analyse due to the sheer quantity of relevant parts and their interrelations, it is not fundamentally impossible to do so.

The questions raised by current research in the applied life sciences illustrate the importance of the discussion over these two competing paradigms. Gene editing, CRISPR-Cas and various developments in synthetic biology (such as the aim to synthesise artificial cells) have made the discussion on the nature of life and the nature of technology highly relevant again. This is specifically the case for the debate on vitalism. The idea of a mystic (for some vitalists a divine) spark of life distinguishing organics from inorganics in biology, is not taken seriously in current molecular life sciences, but whether this is indeed the nail in the coffin for *any* theory that distinguishes between complex living systems and systems that are not alive remains unclear. In 2007, the prestigious scientific journal *Nature* also engaged in this debate with an editorial (*Nature*, 28 June 2007) under the heading 'Meanings of life'. It appears to have been triggered by responses of Pat Mooney, director of the ECT group, to Craig Venter's research (van den Belt 2009), suggesting that Synthetic Biology now competes with God over the design of life. Synthetic biology is paired against the few vitalist claims still persisting in biology. Therefore, the value sensitivities involved in the 'creation of life' again become a central aspect of discussions on biotechnology. In synthetic biology life is explained in terms of molecular processes. Metaphors and archetypes such as Frankenstein's monster or the Golem are thus deemed misguiding and therefore irrelevant to the debate. Still, the definition of life remains up for debate.

Some scholars believe that synthetic biology has put an end to vitalism once and for all, since it demonstrates how organic life is based on inorganic materials, and can be functionally reduced to these factors (Church and Regis 2012). The formation of living structures is thus claimed to be 'gradual, contingent and precarious' (van den Belt 2009). Some scientists approach the issue from a different view on evolution (Westerhoff et al. 2014). They do take into account the complexity of converging and diverging structures as a property intrinsic to life. Both positions bear witness to two different worldviews: the one tending towards mechanistic accounts of life, the other tending towards process-oriented views on life. The former is currently still the most dominant in common life-scientific discourse.

In explanations to both students and wider society over their work, many scientists involved in the engineering aspects of life sciences research refer to their expertise in terms of bio-engineering. This entails the idea that one can approach the basic compounds of a cell in the same fashion as the basic compounds of, say, a car. In essence, the mechanistic worldview is thus again applied to phenomena of life, in the natural world, but specifically for the purpose of mimicking functions of living systems (for instance, the production of specific enzymes) in industrial processes. In this retake of the machine metaphor for life is now applied to its most basic structures. The codification of the basic structure of DNA in terms of C, T, A and G has strengthened the view of DNA in terms of a code or language. A language that first needed to be deciphered to enable us to read it (genomics) but as we succeeded in doing so, we are now developing tools to edit it (Zwart 2012).

The terms biotechnology and bioengineering suggest that progress has been made from describing and understanding the mechanics of life ("analysis") towards becoming the engineer of life ('re-synthesis'). In other words, the study of the mechanics of life has led to the implementation of this mechanics in practice. This move places the life scientist in the position of a car mechanic, who has found the parts with which to build or refashion his car, for example by using components found in the junkyard next to his workplace. Even the metaphorical use of terms refer to such contexts: the term 'chassis' is applied as a standard in synthetic biology research to refer to particular organisms the genetic properties of which can be modified to serve as a basis for multiple uses and functions.[12] Nature offers elements and properties that are only now understood sufficiently to be able to make life-like things. This rationale, as many bioengineers will emphasise is, however, the reverse of what they actually do in practice. The adage that with understanding one can make applications is not valid in this area: life scientists usually do not really understand what happens when they succeed in doing what they do; they rather appear to be successful in what they attempt without fully grasping the underlying principles of the systems they intervene in.

Synthetic biology is an emergent branch of bioengineering that attempts the creation of new life forms. It holds a claim to one of the oldest dreams of science: the creation of life from non-living matter. Already in ancient times, this dream gave rise to a specific branch of study: alchemy. The basic objective of alchemy, now considered to be pseudoscience, was to create the so-called 'philosopher's stone', but it is not very clear what they exactly meant by that. Alchemy uses chemical processes as metaphors for a method to achieve spiritual enlightenment. It was precisely this connection between natural science and psychology that triggered the aforementioned Jung to study the field. Allegedly, the philosopher's stone would enable its owner to create life or turn basic matter into gold. Essentially, this alchemic goal persists in modern synthetic biology: synthetic biology also uses chemical processes for more than mere experiment: in its slipstream, it offers the potential to change not just microorganisms but also the fundaments of human nature.

Creating life from scratch is the golden promise and core objective for ancient, medieval and early modern alchemists. It is also one of the main objectives of synthetic biology: to step beyond mere genetic modification towards the construction of a synthetic microorganism from basic inorganic components. As Bensaude-Vincent (2009) states, the minimal genome approach that lies at the basis of the synthetic biological endeavour to construct new organisms only uses the most basic elements required for sustaining life. The synthetic cell would be a minimal cell, reduced to its core essence, an endeavour that can be compared to the strategy used by alchemist in their attempts at making gold by reducing metals to their primal, pristine state (in Latin: 'reductio in pristinum statum'), thus recovering the potential of matter to become all metals (Bensaude-Vincent 2009). According to alchemy, metals

---

[12] For an example, please see the use of the concept of a safe 'chassis', a standardised GM organism that has multiple safeguards built into its DNA.

also change (from lead into gold for instance) in nature, but this was regarded as an extremely time-consuming process, requiring centuries of centuries of time. The purpose of alchemy was to find a way to speed-up the process (Zwart 2010). One could argue that something similar applies to synthetic biology as well. Life may under certain conditions evolve from inorganic nature, and the idea is now to control and accelerate this process. Tension areas can be found in the self-definition of synthetic biology and in the emerging societal debates on synthetic biology and its objectives. It also raises more philosophical questions, such as: 'will an artificial minimal cell be alive, or will it rather be a fake cell?' (Zwart 2019).

Going beyond mere genetic modification of organisms, synthetic biology attempts to create new organisms from standardised 'lego-pieces' of basic biomaterials ("bio-bricks"). But the lego-metaphor does not cover all aspects. Rather, synthetic biology is an interaction or mutual interpenetration of microbiology and the bio-brick approach, so that bacteria are basically considered as collaborating micro-engineers. The micro-organisms involved are stated to yield the possibility of new approaches to autoimmune and viral diseases such as HIV, to the sustainable production of food, medicine, and to industrial processes such as the production of bio-materials and the management of waste, plastics in particular.

Craig Venter is perhaps the most prominent representative of the reductionist position taken in synthetic biology. He aims to reveal and consciously employ the basic fabrics of life. He refers to the field of synthetic biology as the field that uses the principles of engineering to synthesise new life forms or new functions in existing life forms, so that these can function as tools (Venter 2013). He expects synthetic biology to support the production of clean energy, the resolution of the world food problem and the decontamination of polluted ecosystems. It does so on the basis of a reductionist approach. In an editorial written in 2007, Nature claimed that with synthetic biology all living structures can be reduced to their functional operationalisers (Nature 2007). This triumph of reductionism is a reductionism from the life sciences to organic chemistry, and ultimately biophysics. But the triumph consists of a hidden selfreferentiality: if one studies life as reducible to material composers, one ignores the matter of life as such. To elucidate this point, we should return to the relevance of Bergson's views.

As mentioned earlier, Bergson's gambit, crucial to the writing of *Creative Evolution* was to go beyond either finalism (teleological views on the nature of life) as well as mechanistic thought, that life is reducible to nonlife. The wager of this publication was, therefore, not the same as the wagers surrounding the question of life now: whilst the definition and limitations of life still revolve around the issue of mechanistic reductionism, the problem now concerns the tension between mechanistic determinism on the one hand and the nature of complex systems; on the other not the expectation that life serves some kind of pregiven goal – be it defined in Aristotelian or theological terms. The reading of the following paragraphs, in their quoting Bergson, needs to be guided by an awareness of this changed discourse.

In *Creative* Evolution Bergson stated: "[…] histologists, embryogenists, and naturalists believe far less readily than physiologists in the physico-chemical character of vital actions. The fact is, neither one nor the other of these two theories, neither that which affirms nor that which denies the possibility of chemically producing an elementary organism, can claim the authority of experiment. They are both unverifiable, the former because science has not yet advanced a step toward the chemical synthesis of a living substance, the second because there is no conceivable way of proving experimentally the impossibility of a fact. But we have set forth the theoretical reasons which prevent us from likening the living being, a system closed off by nature, to the systems which our science isolates." (Bergson 1911 [1907], pp. 36). In 1908, the year of the original publication of *Creative Evolution*, the chemical synthesis of life was still a far way off. But already then, Bergson claims that life and nonlife should not be likened. Bergson went on to write: "The mechanistic explanations, we said, hold good for the systems that our thought artificially detaches from the whole. But of the whole itself and of the systems which, within this whole, seem to take after it, we cannot admit a priori that they are mechanically explicable, for then time would be useless, and even unreal. The essence of mechanical explanation, in fact, is to regard the future and the past as calculable functions of the present, and thus to claim that all is given." (Bergson 1911 [1907], p. 37). As such, they cannot explain the phenomenon of life, since life, as we will see later on, is fundamentally innovative. And as a result "[...]analysis will undoubtedly resolve the process of organic creation into an ever-growing number of physico-chemical phenomena, and chemists and physicists will have to do, of course, with nothing but these. But it does not follow that chemistry and physics will ever give us the key to life."

Synthetic biology embraces an engineering view on life. Analysis will undoubtedly resolve the process of organic creation into an ever-growing number of physico-chemical phenomena, and chemists and physicists will have to do, of course, with nothing but these. But it does not follow that chemistry and physics will ever give us the key to life. But although the engineering approach may be highly successful when it comes to modifying organisms to engineer functional products, this does not mean that it allows us to understand what life really is. The successes of bioengineering are enormous, spectacular perhaps, but, as Bergson argues, they merely involve those aspects of life (for instance, those aspects of the functioning of a cell) that are mechanical and repetitive in the first place, it does not address the more creative and free-floating dimensions of living processes. It merely excludes these from the territory of study. It does so legitimately since it holds different and more pragmatic goals, but its conclusions should not be drawn beyond these goals.

I briefly referred to complex systems theory earlier. In recent years, complexity theory has gained importance in various disciplines, in physics, but also in chemistry, biology, organisational psychology and sociology. Complexity theory aims to elucidate properties that are typical for complex systems. It attempts to explain the behaviour of a system the components of which demonstrate such a multitude of interactions that an explanation of the system's behaviour cannot be reduced to an account of the behaviour of its components (Cath 2018). Since there is no reasonable way to track down all possible interactions and deduce their collective sum, complexity theory requires a systems approach. It provides for an account of complex systems behaviour on the basis of a holistic approach, in which behaviour is not caused but emerges through complex interactions. The system as such is self-organising rather than being determined by external factors.

Synthetic biology is often associated with complex systems theory: it not only builds on notions derived from systems biology to understand complex systems behaviour, it also considers big data as part of its methodological portfolio – an approach in which informational patterns become an important source of insight, provided one has sufficiently large data sets at one's disposal. On closer inspection, however, it appears that, notwithstanding this openness to a systems approach, the basic tendencies of synthetic biology remain closely related to the mechanistic way of thinking (Gharud et al. 2015, p. 37). The conflict involved is still the conflict between an organic and a mechanistic worldview. Here, it becomes apparent that the engineering perspective in biotechnology, although useful in many respects, does not deliver a ultimate answer to the question of life. It merely focuses on a limited number of functionalisable aspects of living organisms. Bergson urges cautious scepticism about the scientific ability to reduce higher organisms to mechanical functions, although admitting this might be possible for amoebae. Here, his caution was proven to be right. But his critique of reductionism still remains relevant.

The concepts of spontaneously emerging properties, converging and diverging systems, complexity etc. belong to the first worldview whilst concepts such as building blocks, cell synthesis, function etc. are part of a vocabulary associated with the second. They represent two traditions in the history of biology, the one involving systems thinkers – focusing on the complexity and history of living systems (and their interrelation), the other involving physiologists – focusing on the biochemical aspects of life. As Bergson writes in *Creative Evolution*:

> "Those whose attention is concentrated on the minute structure of living tissues, on their genesis and evolution, histologists and embryogenists on the one hand, naturalists on the other, are interested in the retort itself, not merely in its contents. They find that this retort creates its own form through a unique series of acts that really constitute a history. Thus, histologists, embryogenists, and naturalists believe far less readily than physiologists in the physico-chemical character of vital actions." (Bergson 1911 [1907], p. 51).

Referring to the superhuman intellect, later dubbed 'the demon of Laplace' (discussed in paragraph 3.1) Bergson explains that the problem of a radical mechanistic worldview lies in its implicit embrace of a metaphysics in which everything real is imagined to be given, laid out on the table of eternity. And the fact that some phenomena cannot be predicted, is attributed to the "infirmity of a mind that cannot know everything at once" (Bergson 1911 [1907]). Bergson instead believed that our immediate experience of life as a creative process should inform our conclusions. Bergson explained this in his *Introduction to Metaphysics* (1903) with the following phrase: "To analyze, therefore, is to express a thing as a function of something other than itself. It is the analytic mind that stands in its own way, but intuition can go further. And here, we can only proceed if we acknowledge duration" (Bergson 1999 [1903]), which might be understood as change in terms of irreversible and self-organising processes of divergence and convergence.

Evolution theory derives its value from its self-evidence. Accidental, spontaneous variation lies at the basis of evolutionary processes: many of the mutations and variations that occur are traditionally regarded as 'white noise' – an annoying background noise to the musical composition of life. In this view, variations that appear as problematic to survival within a given ecosystem are often seen as debris accumulated during the supposedly inefficient process of evolution. It is, however, this specific 'white noise' that lies at the heart of evolutionary processes (Clark 2010; Westerhoff et al. 2014). Moreover, the notion of efficiency implicitly presupposes the notion of telos (purpose). The spontaneous emergence of new variants and new species is therefore necessarily intertwined with this aspect. The entropic nature of variability is, paradoxically perhaps, the basis for the negentropic (Schrödinger 1944), self-organising nature of evolution. The ability to use the material basis of the complex processes of evolution for engineering purposes does not equal an understanding of these complex purposes. But although even the primitive amoeba is much more complex than Bergson believed, there is, in his view, boundary between life and nonlife. It is merely typified by a strive for individuality: "individuality is never perfect, and that it is often difficult, sometimes impossible, to tell what is an individual, and what is not, but that life nevertheless manifests a search for individuality, as if it strove to constitute systems naturally isolated, naturally closed." (Bergson 1911 [1907]). In other words "certain aspects of the present, important for science, are calculable as functions of the immediate past. Nothing of the sort in the domain of life. Here calculation touches, at most, certain phenomena of organic destruction. Organic creation, on the contrary, the evolutionary phenomena which properly constitute life, we cannot in any way subject to a mathematical treatment." (Bergson 1911 [1907], pp. 20). Now, one might ask oneself whether it is not hypothetically possible to explain life in the same deterministic way as the movements of the solar system, but, apart from the 'superhuman calculator', in Bergson's phrasing, that one needs to imagine to be able to conduct the task, it is a tendency of our natural metaphysical inclinations to imagine such determinism.

As Bergson poses "Any attempt to distinguish between an artificial and a natural system, between the dead and the living, runs counter to this tendency at once. Thus it happens that we find it equally difficult to imagine that the organized has duration and that the unorganized has not." (Bergson 1911 [1907], p. 21). The relevance for the debate on synthetic biology could not be greater. Since here, Bergson alludes to the difficulty to set apart the realms of life from the rest of reality. And indeed, why would one want to set apart life in the first place? Only non-scientific motifs seem to inform such a position. But no, here it needs to be clarified where Bergson draws the distinction. Whilst biological vitalists of the nineteenth century drew the distinction between life and nonlife at the ontic level – speaking as it were of two different substances, Bergson draws the distinction in relation to duration, in relation to how life manifests a concrete creativity, whilst nonlife does not. It is therefore invention (or self-innovation if you prefer) that distinguishes life from nonlife. Referring to the genetic experiments of the Dutch biologist Hugo de Vries (1848–1935) he considered the emergent science of genetics as confirming rather than denying this property of creativity in life. But if it is so distinct from nonlife, how could it have ever emerged from it? After all, it is the unexplained origin of life from nonlife in which biological vitalist thesis found its main argument. Bergson remains somewhat vague about this: "At a certain moment, in certain points of space, a visible current has taken rise; this current of life, traversing the bodies it has organized one after another, passing from generation to generation, has become divided amongst species and distributed amongst individuals without losing anything of its force, rather intensifying in proportion to its advance." (Bergson 1911 [1907], p. 26). This certain moment is connected to what Bergson refers to as 'supraconsciousness' or a 'need of creation' that is dormant when life is condemned to automatism (Bergson 1911 [1907], p. 261). To be able to detail this possibly somewhat cryptic position, we may need to briefly take a sidestep, and lend support for this position from the field itself.[13]

Here it might be enlightening to draw a parallel to the work of the early immunologist Elie Metchnikoff. Although Bergson and Metchnikoff did meet, they only met informally and only discussed an indirectly related subject matter at that occasion – namely immortality. The elaboration below outlines the terms for a dialogue that never happened, a missed encounter.

Metchnikoff was a contemporary of Bergson. Elie Metchnikoff defined the basic properties of life as it emerges through his definition of the immune system (Mutsaers 2016): the immune system is the very mechanism that defines the individuality of an organism. This implies that immunity is what basically defines the organism as such. Translated into the language of organic chemistry of today: although the coming into existence of complex organic molecules still is speculated to have originated through lightning, the transition to complex carbon-based molecules to life forms that are able to persist and reproduce does not occurs under the influence of some kind of spark. It is rather the case that, where such complex molecules evolved, an immune function emerged, thus distinguishing between what was welcome to the system and what not. In this manner, a life-like entity becomes an organism proper. The evolution from complex carbon-based molecules to self-replicating organisms came about because those systems that did not accidentally self-replicate did not persist, whilst some apparently did. But without the self-identifying and other-identifying system of immunity, there are no individual organisms. There is no self before the definition of what is proper to that organism and what is regarded as foreign. These insights delegated a central role to the immune system in the emergence of life, a dynamic account of the emergence of 'self' and 'other'. This view, now considered defunct because Phylum Protista, (Amoeba, Hydra, Euglena etc.) are life forms without the ability to define self in relation to non-self via an immune system – something which is also true for many bacterial types (ref.).

Metchnikoff once attended a public lecture given by Henri Bergson, and, being critical of the tendency of his contemporaries to accept intuitionist anti-rationalism, was concerned not so much by the ideas of Bergson, but rather by the mystical prejudice of the audiences typically attending such lectures (Vikhanski 2016). At

---

[13] And admittedly, by adhering to a position that, although articulated by a contemporary of Bergson, was not picked up by him in spite of its crucial importance to his position.

the time, the younger generation was critical of positivism, looking for answers in metaphysics instead (Stambler 2014). Metchnikoff: "Bergson preaches the limitations of knowledge and valorises intuition, struggling to convince us of the existence of a soul independent of the brain function and of the existence of a free will."(Stambler 2014). Metchnikoff himself, however, hoped to answer the question of immortality on a scientific rather than a metaphysical basis. Bergson's *Creative Evolution* touched upon several aspects of Metchnikoff's *The Prolongation of Life: Optimistic Studies* was also translated but unfortunately, his views on the origin and self-defining nature of life were not touched upon in any formal publication of Bergson.[14]

Metchnikoff places the immune system at the heart of the origin of life. For him, it was the very defining mechanism for the self-identification of life as such. And in this regard his views explain the nature of the origin of life and evolution. It is a central argument against vitalism in biology that does not straightforwardly dismiss with the distinction between life and nonlife. Metchnikoff never used the metaphor of the immune system as a defensive army of the organism against external threats. Metchnikoff's initial ideas were in part misrepresented in later immunology discourse, where the immune system was defined as a defence mechanism of an organism against alien invasions. This military metaphor for the immune function obfuscated the relevance of the initial view for explaining the origin of life. This obfuscation may have been caused by Metchnikoff's use of the term immunology as such, a term that originated in the realm of political discourse, rather than in biology (Mutsaers 2016).

The immunological theories of Mechnikoff form an important hypothesis of how life originally developed from non-life – the proverbial primordial soup. Metchnikoff's views on immunology may help to explain how life evolved from non-living (self-reproducing) molecules into multi-cellular organisms (plants, fungi and animals, the most primitive of which are the sponges). His theories explain the birth of life in terms of the emergence of the ability to self-identify: this is what defines the organism as organism. It has an ability to have a self since it has been able to individuate from its surroundings. In other words, life, organism and self cannot be separated. It explains how life emerged from this primordial soup, in which complex molecules managed to survive longer than might be expected, from a merely chemical point of view: only at another level of aggregation can these complex autopoietic tendencies or properties be accounted for.

Metchnikoff's work lies at the historical origins of modern immunology. His views on the role of the immune system still form a valuable contribution to the debate on biological vitalism. Rather than portraying the immune system as an army to protect the organism against enemy attacks, he perceived of it as the very system that distinguishes self from other. With this view a different picture emerges concerning the boundaries of the self. The organism is not identified by some kind of

---

[14] Since much of Bergson's personal archive was destroyed at his own request, we cannot know whether he might have reflected upon the potential of Metchnikoff's theories more elaborately.

externally given special nature that define it as different from the inorganic. It is defined by its ability to identify between self and other, and this ability is the immune system. Metchnikoff developed an idea over the emergence of life in which it is the immune system that distinguishes self from other. This delimitation should be regarded as the most basic principle of life.

If we follow Metchnikoff's account of the origin of life, we can manage to transcend the biological vitalist distinction between living and non-living, between organic and inorganic. Although hardly recognised as such, Metchninkoff provided an important contribution to the abiogenic study of the origins of evolution. It explains the step which must have occurred after the emergence of complex carbon molecules. The teleological view on the temporal aspect of evolution is a remnant of object-metaphysics. It takes species as object-categories, and still ranks them along the taxonomic principles of the eighteenth century, whilst neglecting the fact that the spaces between species are fluid. It also perceives of species and individuals as either functional or dysfunctional (adapting or not-adapting) within such a teleology of increasing progress. And it interprets the microbiological principles of variation as mechanisms.

Object metaphysics perceives of reality in terms of a collection of things, made up of smaller things. In essence, it is an atomistic worldview. In the life sciences, it is articulated in the view that organisms are made up of cells, and cells of again smaller components, the most central one being DNA. DNA can again be read as a molecule made up of smaller units (C, T, A, G) that function as the hardware or blueprint for a cell's functioning, and thus an organism's functioning. The cell is regarded as a machine.

In the life sciences, the metaphor of DNA as the alphabet of life replaces the machine metaphor. The pre-existing metaphor of organisms as machines or automatons – dating back to René Descartes – gives way to the metaphor of the computer, while more recently, 'reading' the code of life gives way to the idea of '(re)writing' the code of life (Craig Venter; Cf. Zwart 2012). The current revolution in the life sciences is moving beyond the machine metaphor in the traditional mechanistic sense. Living entities cannot be compared to an engine such as a car's. Rather, the basic components of life function like an alphabet. Metaphors such as the clock or the steam engine are now replaced by the metaphor of the computer: a completely different, post-mechanistic kind of machine, based on information.

The transition from classical biology via microbiology up to synthetic biology reflects, philosophically speaking, a development from mechanistic thinking to process thinking. Whilst we have been comparing biological processes to the intricate mechanics that went into our clockworks, engines and eventually computers, the technological evolution from clockworks via engines to computers actually bridges the gap between living entities and machines, because computer-like contrivances are much more similar to living entities than clock-works are. This also means that we can understand the technologies which we currently generate in terms of living systems, namely as converging and evolving processes and networks.

Biotechnology exemplifies contemporary technoscience because it combines biological research with engineering. In the field of biotechnology, biology and engineering collaborate to the extent of merging into one discipline (bioengineering), to generate innovative products. In earlier phases, a rhetoric was dominant in biotechnology which still seemed to reflect the atomist worldview: genes were presented as stable bits of information, without taking issues of context into account that decide whether genes are active or not, and how they can self-modify. Currently, however, a systems or process thinking approach is becoming increasingly dominant in biotechnology. Bioengineering principles are applied to biological systems, and this should not be seen as a machine-approach to life, but rather as a life-approach to machinery. Technology and human existence are becoming deeply intertwined, the evolution of humans and of technology is converging: a creative process which unfolds in time.

For the modern life sciences, it is attractive to resolve the dualism between the paradigm of object metaphysics and the paradigm of process thinking by reducing the processes of life to the basic nucleic components. We may, however, also resolve this dualism by emphasising basic aspects of evolution discernible in non-living systems such as change, complexity etc. In this regard, the potential for evolutionary development is omnipresent, and also rudimentary manifest in non-living natural phenomena, from the birth of stars down to the growth of crystals. Life merely represents a higher aggregation level of complexity, where self-organising processes have emerged.

Again, Bergson's élan vital is not the same as nineteenth century Berzelian vitalism: it does not involve a divine spark of life. It is the manifestation of time, in its concrete shape, as the ever-evolving nature of life. Life, in this perspective, is merely a synonym for change, in its self-organising complexity. The issue of the origin of life can only be successfully addressed if we acknowledge the problem of the freezing of time, the tendency of the applied life sciences to reduce the very phenomenon of life to predictable processes taking place on the scale of clock-time, rather than as unpredictable, creative processes occurring in the course of duration. Although the focus on small-scale in vitro processes seem legitimate from a pragmatic laboratory perspective, we can only really understand evolution if we de-freeze our conception of time, focusing on auto-poetic or self-creative emergence. From a Bergsonian duration perspective, 'white noise' – the apparent debris of evolution – becomes a key aspect of the evolutionary processes of life (Westerhoff et al. 2014).

The computer metaphor is not restricted to the life sciences. Similar problems seem to affect the study of the human mind. The metaphor of the computer still functions as a basic metaphor for how the mind works. Our brain is the hardware, our memories, linguistic functioning, emotional experiences and creativity are the software. In both neuroscience and modern psychology this remains a dominant metaphor. The metaphor however holds several problems, dependent on whether the computer-metaphor is used in a reductionistic or rather in the systems sense. In the reductionistic version, the status of knowledge becomes reduced to arbitrary data stored on a pregiven hard drive. The mind would then function like a pre-programmed player piano, biologically conditioned to play the tunes fed to us through perforated paper or metal cartridges. But this mechanistic view on the human mind sees music as a pre-programmed tune, perceived by a pre-programmed audience. In reality, however, playing the piano is a much more creative programme, and there are degrees of freedom or creativity between the score and the performance. For the philosopher of science, this creates a kaleidoscopic problem: each time we see our own neurobiological nature as pre-programmed, we understand the functioning of our mind as merely a programme. This leads us back to the mechanistic idea of a programme that was written by a divine being, feeding us with the cartridges. For Bergson, this mechanistic container-view of the human mind is problematic, and he criticizes it, especially in the psychological literature of his time. In the next chapter, I will give a more detailed account of Bergson's criticism of mechanistic interpretations of the human mind.

Literature

Bensaude-Vincent, B. (2009). Biomimetic chemistry and synthetic biology: A two-way traffic across the Borders. *International Journal for Philosophy of Chemistry, 15*(1), 31–46.

Bergson, H. (1912 [1903]). *An introduction to metaphysics* (trans; Hulme, T.E.). Hackett Publishing Company 1999. Also included in: Bergson, H. (2007 (1923)). *The creative mind: An introduction to metaphysics* 1923. Dover Publications: Dover.

Bergson, H. (1911 [1907]). *Creative evolution*. New York: Henry Holt and Company.

Bohm, D. (1980). *Wholeness and the implicate order*. London: Routledge.

Brillouin, L. ([1956, 1962] 2004). *Science and information theory*. Mineola, NY: Dover.

Broeks, D., & Zwart, H. (2021, forthcoming). Understanding life by engineering it: An epistemological assessment of the Feynman-dictum in synthetic biology. *History and Philosophy of the Life Sciences*.

Cath, A. G. (2018). *Solace in complexity: A fundamental empirical thought experiment*. Vianen: Uitgeverij ProefschriftMaken.

Church, G., & Regis E. (2012). *Regenesis. How synthetic biology will reinvent nature and ourselves*. Philadelphia: Basic Books.

Clark, K. B. (2010). Origins of learned reciprocity in solitary ciliates searching grouped courting assurances at quantum efficiencies. *Biosystems, 99*, 27–41. https://doi.org/10.1016/j.biosystems.2009.08.005.

Crick, F. H., & Orgel, L. E. (1973). Directed panspermia. *Icarus, 19*(3), 341–348.

Crow, J. F. (1992). Erwin Schrödinger and the hornless cattle problem. *Genetics, 130*, 237–239.

Evans, F. C. (1951). Ecology and urban areal research. *Scientific Monthly, 73*.

Gharud, R., Simpson, B., Langley, A., & Tsoukas, H. (2015). *The emergence of novelty in organisations*. Oxford: Oxford University Press.

Gould, S. J. (1995). *What is life? The next fifty years: Speculations on the future of biology*. Cambridge: Cambridge University Press.

Kant, I. (1781). *Kritik der reinen Vernunft (A)*. Riga: Hartknoch.

Miquel, P. (2007). Bergson and Darwin: From an Immanentist to an Emergentist approach to evolution. *SubStance, 36*(3), 42–56. Issue 114: Henri Bergson's creative evolution 100 years later (2007). Published by: University of Wisconsin Press.

Morin, E. (1990). *Introduction a la pensee complexe*. Paris: ESF.

Mutsaers, I. (2016). *Immunological discourse in political philosophy: Immunisation and its discontents*. London: Routledge.

Nature. (2007). Nature editorial: Meanings of life: Synthetic biology provides a welcome antidote to chronic vitalism. *Nature, 447*, 1031–1032.

Onffray de Lamettrie, J. (2003 [1748]). *Machine man and other writings* (ed: Thomson, A.). Cambridge: Cambridge University Press.

Schrödinger, E. (1944). *What is life – The physical aspect of the living cell*. Cambridge: Cambridge University Press.

Spencer, H. (1854). *Essays: Scientific, political and speculative* (Vol. 2). London: Williams and Norgate.

Spier, R. E. (2003). History of animal cell technology. In R. E. Spier (Ed.), *Encyclopedia of cell technology* (Vol. 2, pp. 853–872). New York: Wiley.

Stambler I. (2014). A history of life-extensionism in the twentieth century. *Longevity History, Rishon Lezion*. http://www.longevityhistory.com. Last accessed: 10-07-2019.

van den Belt, H. (2009). Playing god in Frankenstein's footsteps: Synthetic biology and the meaning of life. *NanoEthics, 3*(3), 257–268. https://doi.org/10.1007/s11569-009-0079-6.

Venter, C. (2013). *Life at the speed of light: From the double Helix to the Dawn of digital life*. New York: Viking Publishing.

Vikhanski, L. (2016). *Immunity: How Elie Metchnikoff changed the course of modern medicine*. Chicago: Chicago Review Press.

von Bertalanffy, L. (1934). Untersuchungen über die Gesetzlichkeit des Wachstums I: Allgemeine Grundlagen der Theorie; mathematische und physiologische Gesetzlichkeiten des Wachstums bei Wassertieren. *Arch. Entwicklungsmech, 131*, 613–652.

Westerhoff, H. V., Brooks, A. N., Simeonidis, E., García-Contreras, R., He, F., Boogerd, F. C., et al. (2014). Macromolecular networks and intelligence in microorganisms. *Frontiers in Microbiology, 5*(379).

Wickramasinghe, C. (2011). Bacterial morphologies supporting cometary panspermia: A reappraisal. *International Journal of Astrobiology, 10*(1), 25–30.

Zwart, H. (2010). *De waarheid op de wand: psychoanalyse van het weten*. Nijmegen: Vantilt. ISBN 9789460040481.

Zwart, H. (2012). On decoding and rewriting genomes: A psychoanalytical reading of a scientific revolution. *Medicine, Healthcare and Philosophy: A European Journal, 15*(3), 337–346.

Zwart, H. (2018). In the beginning was the genome: Genomics and the bi-textuality of human existence. *The New Bioethics: A Multidisciplinary Journal of Biotechnology and the Body*. https://doi.org/10.1080/20502877.2018.1438776.

Zwart, H. (2019). What is mimicked by biomimicry? Synthetic cells as exemplifications of the three-fold biomimicry paradox. *Environmental Values, 27*(special issue).

Chapter 5

Time, Life and Memory: Bergson and Neuroscience

> [W]hether, indeed, thought is regarded as a mere function of the brain and the state of consciousness as an epiphenomenon of the state of the brain, or whether mental states and brain states are held to be two versions, in two different languages, of one and the same original, in either case it is laid down that, could we penetrate into the inside of a brain at work and behold the dance of the atoms which make up the cortex, and if, on the other hand, we possessed the key to psycho-physiology, we should know every detail of what is going on in the corresponding consciousness. Henri Bergson, *Matter and Memory*

The associations triggered by the term 'mind' tend to be very different from those associated with the term 'brain'. The brain is associated with the anatomical substance found inside our cranium. If the top of a human skull is lifted, as a lid on a jar - as in anatomical lessons (painted by Rembrandt for instance) – what is revealed is a brain, rather than a mind. We believe this is the anatomic location where processes that we experience as 'our mind' can be found. Some were led to believe that human abnormalities could be traced in the shape under that lid: too big, too small, weird lumps etc. In this vein, the brain is an object. The concept of the mind is associated with the subjective experience of a stream of consciousness. It belongs with a paradigm of self-aware conscious thought. The mind can be reflexive. It is able to bend back upon itself: when we are a mind thinking about a mind, that is: thinking about ourselves. It is this theme that is addressed by the philosophy of mind.[1]

Whereas the brain can be anatomically dissected, the concept of the mind rather involves introspection. Whereas the brain is an object of science, the mind is a more or less phenomenological concept. It is not something which *appears to us*, but rather *that to which other things (such as brains) appear*. In epistemological terms, the problem is that, although brain and mind are somehow related, the brain is an *object*, while the mind is connected with the position of the *subject*. It is this special status of the mind, or of consciousness, which triggered Bergson to develop his

---

[1] In the case of the heart, things are different again. It is both an anatomical concept and part of a different vocabulary, associated the experienced with the sacred, with compassion, empathy and love. The concept of the soul functions within a religious context. It is taken to be that aspect of our being that stands in the face of eternity, carrying our whole being, but not something whose location can be anatomically determined.

© Springer Nature Switzerland AG 2021
L. Landeweerd, *Time, Life & Memory*, Library of Ethics and Applied Philosophy 38, https://doi.org/10.1007/978-3-030-56853-5_5

ideas on memory, although the concept of memory is also closely connected with the concept of duration.

For Bergson, the main reason to address the nature of memory was that the phenomenon created a problem for his ideas on duration. We experience of a rupture between 'the past' and 'the present moment' whilst time, as duration, should be thought of in terms of an 'indivisible continuity': without ruptures. To address this apparent contradiction, he needed to tackle an important issue, a legacy more or less[2] of the philosophy of René Descartes: the dualistic separation between mind and body. Bergson aimed to avoid the problematic inheritance of Cartesian dualism. For Bergson, memory was of a spiritual rather than a material nature. Thus, it should not be reduced to merely a property of the brain. He expressed his views in *Matter and Memory* (1896). *Matter and Memory,* (originally published as *Matière et mémoire: Essai sur la relation du corps avec l'esprit* and translated into English in 1908).

In *Matter and Memory*, Bergson develops an alternative to the age-old mind-body distinction and the ontological problems that this distinction carries along – more specifically: the deadlock in which proponents of materialism and idealism ended up during the nineteenth century. At the same time, it aimed to delimit the scientific understanding of the brain from metaphysics.[3] For Bergson, the issue of mind versus matter cannot be resolved by purely looking at empirical findings: the distinction in question is categorical in nature, and therefore entails the danger of confusion between two perspectives.

The idea that our thoughts, emotions and perceptions can be located in the brain remains quite dominant. Anything we associate with mind, spirit, soul or heart is supposed to be located somewhere in this spherical organ located behind our eyes. Terms such as mind, brain, soul and heart reflect incommensurable metaphysical positions, and yet they are all somehow interrelated, because they are all somehow associated with the brain. We (our brains) are part of this world and at the same time we (our mins) are witnessing the phenomena that unfold in this world. Furthermore, we are also witnessing events that take place in our own mind. What we are remains a puzzle. Both world and self constantly reveal themselves, and always in different ways: they are never the same but constantly changing.

From a neuroreductionist perspective, it would seem that we can indeed know what we are by knowing what unfolds in the spherical organ that is supposed to contain our consciousness. What we are – mind, spirit, soul – seems reducible to the neurophysiological structures and processes of the brain. But in mapping the processes of the brain, we at the same time distance ourselves from what we are: this

---

[2] Descartes philosophy is often presented in a more simplistic way than it deserves.

[3] In Bergson's view, metaphysics should not be understood as esoteric spiritualism but rather as the area of research that leans on intuition instead of analytic empiricism. As with other scientific domains that Bergson discussed, Bergson deemed this delimitation to be necessary to avoid a confusion between scientific statements and metaphysical claims. The latter concern, amongst others, claims about the presuppositions of science. But all too often, these are mistaken for the conclusions of scientific research.

exercise does not bring us closer to any definitive knowledge about ourselves. As Keith Ansell-Pearson phrases it: "our psychical life, while bound to its motor accompaniment, is not governed by it" (Ansell-Pearson 2018, p. 77).

At the turn of the century, the issue of the distinction between mind and matter (between spirit and body, between the subject that experiences, observes, reflects and the object thereof) was part of a broader debate on the inheritance of Cartesian dualism. Some participants in the debate took a reductionist position. The knowledge subject is reducible to something objective (e.g. the human brain), as are its contents (thoughts, emotions, perceptions, etc., accessible via empirical studies). There is something unsatisfactory or inconsistent about this position, however. Can we say that memory is merely a brain function (a research object), while at the same time experiencing memory as something which is part of the subject (the one doing the research)? On the one hand, the brain appears within a particular perspective (say, neurophysiology), while at the same time we (our minds) can critically reflect on the validity and limits of this perspective.

For Bergson, our minds cannot be regarded a mere medium to store data. To put it in twenty-first century terms, it is not some kind of biomechanical hard drive to store past experiences. Memory, to him, is intimately connected with duration – lived and experienced time. With this focus on the temporal aspect of the world – on duration – Bergson aimed to avoid a substantivist definition of the relation between mind and matter, spirit and brain. The issue of the substantivist definition of the mind (as some type of spiritual thing, a *res cogitans*) had been one of the most tenacious complications of Cartesian dualism. Many critics of Descartes therefore opted for some kind of monism, either by reducing everything to matter (materialism) or by reducing everything to spirit or mind (idealism). Bergson tried to define an alternative that avoids all three positions: classic dualism, materialist reductionism and idealism.

We appear to wind and rewind between the two poles of our existence (mind and body, matter and mind). The separation between the two was phrased by Descartes as a distinction between two different substances. The idea that the mind is a substance without matter comes across rather uncomfortably. Some scholars (Edmund Husserl specifically) therefore posed that we should regard the distinction between the two as a matter of perspective. Translated into discursive terms, mind would then be the perspective of the first tense (me, myself, I) and matter that which appears within that perspective and is referred to in the third tense (it, that). But during the nineteenth century, the Cartesian distinction between mind and matter still dominated the academic debate.

As regards the relationship between mind and body, the doctrine that dominated the nineteenth century was psychophysiological parallelism. As Bergson wrote in *Mind Energy*: "A superhuman intelligence, watching the dance of the atoms of which the human brain consists and possessing the psycho-physiological key, would be able to read, in the workings of the brain, all that is occurring in the corresponding consciousness" (Bergson , pp. 231–232). This position was already elaborated by the mathematician and philosopher Gottfried Wilhelm (von) Leibniz (1646–1716) and philosopher Nicolas de Malebranche (1638–1715) who termed it 'occasionalism'. But psychophysiological parallelism received renewed attention towards the end of the nineteenth century in discussions over the nature of knowledge in the emerging discipline of psychology. This doctrine tried to avoid the problem suggested by Cartesian dualism that either phenomena of the mind cause phenomena of the body or vice versa. The problem entailed in this problem was that such causation between two different substances could not be explained. Psychophysiological parallelism attempted to circumvent this problem by holding that psychological phenomena merely correlate to physiological phenomena: in other words, a synchrony rather than a cause-effect relationship. However, although its distancing itself from the problematic issue of causation, this view was still in need of further explanation. In his occasionalism, Malebranche had suggested that some type of divine interception mediated between mind and body. Spinoza's solution to the problem was perhaps the most elegant one: he stated that thought and extension (mind and matter) were not to be seen as two separate realities, but rather as two ways of understanding the same thing.

In spite of the problems concerning the logical inconsistency of psychophysiological parallelism, the position remained attractive as a pragmatic solution to the mind-body problem. Both Gustav Theodor Fechner[4] (1801–1887), physicist, philosopher and psychologist, and Wilhelm Max Wundt (1832–1920), founding father of experimental psychology, adhered to this position. Bergson did not consider the pragmatic resolution to be satisfactory, however.

If one does not accept psychophysiological parallelism as a solution to the problems associated with Cartesian dualism, it appears that only two other philosophical positions remain: idealism or materialism. Both have a longstanding tradition, and both carry along their own problems of logical consistency. Before entering into Bergson's analysis of dualism, I here briefly summarise these two schools of thought.

Idealism holds that mind (consciousness, thought) is a first order reality. All that we regard as material is only of a second order: a phenomenon in the mind, rather than an autonomous reality. The English bishop and philosopher George Berkeley (1685–1753) approached empiricism from an idealist position. He held that there was only spirit, and that what we consider to be the real and material world is only what God plants in our perception. In this regard, everything apart from God and ourselves only exists to the extent that it is perceived (Latin: *esse est percipi*). As a result, idealist philosophers have a problem in explaining why real objects would be different from imaginary ones. And when drawn out to the extreme: if the world and everything in it merely exists in our minds, to the extent that we perceive of it, the world is merely a figment of our imagination. This solipsist conclusion is usually not accepted.

Materialism takes an opposite position from idealism. It holds that all that exists is first and foremost physical and material in nature, and as such it adheres to the

---

[4] Bergson devoted his 'Time and Free Will', his doctoral thesis to the critique of Fechner's viewpoints as they were expressed in the Law of Fechner, which defines the parallelism between the actual change in physical stimuli and the perceived changes thereof.

laws of the material world. Whilst modern materialism can be traced back to Julien Offray de Lamettrie (see Chap. 3), it became more articulate in the neo-Kantian school of philosophy. Neo-Kantian materialist philosophers such as Hermann Ludwig Ferdinand von Helmholtz (1821–1894), Karl Vogt (1817–1895), Heinrich Czolbe (1819–1873) and Ludwig Büchner (1824–1899) tended to interpret Kantian epistemology in terms of the psychological structure of the brain. Our knowledge categories (time, space, causation, etc.) should be seen as pregiven due to the structure of our brains, rather than as formal categories of rational necessity. In their minds, the physical processes of the brain provided a steadier point of departure for an account of the a priori conditions of experience. In their view, Kant's transcendental argument, due to its adherence to something that transcends nature, formed a too porous foundation for the architectonics of his epistemology.

Epiphenomenalism is a specific materialist position on the mind–body problem. It considers physiological and biochemical processes (such as sensory perception, neural impulses, the movements of the bowels, the muscles etc.) as causes of mental processes such as thought, emotions, awareness and cognition. These are not to be considered as transcendent. They only appear to be autonomous and real, but in fact the merely exist 'in the eye of the beholder', and they are intrinsically and ultimately material in nature. This means that all subjective experiences (all notions of self, self-awareness, consciousness etc.) are ultimately biochemically determined. This position was specifically dominant in behaviourism (Ivan Pavlov (1849–1936), John B. Watson (1878–1958), and B. F. Skinner (1904–1990)).

For Bergson, the distinction between matter and mind, between the physiological and the mental (or spiritual) should not be defined in terms of spatial concepts (inside versus outside, internal versus external), but on the basis of their temporal properties. It is by this step that Bergson steps out of the Cartesian metaphysical worldview that remained dominant over the course of the nineteenth century. Bergson, critical of the Kantian tradition, does agree with Kant that the mind should not be seen as substance. But Bergson emphasises an understanding of the mind in terms of process, movement, evolution, in short as duration, and not in terms of transcendence.

Immanuel Kant's position regarding the world of phenomena (things as they appear to us, as opposed to the noumenal,[5] the world of things as they exist outside of our sensory perception and outside of our cognitive systems) is that, studied from a scientific point of view, we have to assume that all events are determined by causation. Kant terms this position transcendental idealism; it allows for the existence of

---

[5]The phrasing of 'noumenon' (Greek: νούμενον (noúmenon) originally means 'that which is thought', whilst it has come to refer to that which cannot be thought: that which is outside of the sphere of our perception of the world, the world as it exists outside of our grasp of it – as contrasted with phenomenon (Greek: φαινόμενον (phainómenon)), that which appears to us, reveals itself to us. The two extremes of things as they exist upon themselves and things as they appear to us, meet in our conception of ourselves as noumenon: we ourselves are the only 'thing in itself' to which we, at least in the conception thereof in Kantian epistemology, retain immediate access, but even then, not in an intellectual or cognitive sense. See also footnote 25.

a world in itself, but this world cannot be perceived by us. We necessarily perceive the world in terms of space and time, and how that world exists in itself is beyond our experiential horizon. Causal determinism for Kant is therefore not a property of the world as it exists outside of our perception, but a property of our way of knowing the world. Although not problematic when studying stars, stones or molecules, this does become problematic when studying the nature of consciousness. Then, the epistemological subject all of a sudden becomes the epistemological object. It becomes part of the phenomenal world. Kant therefore defined causal determinism as an aspect of our way of knowing of the world, rather than as an intrinsic feature of that world itself. Neo-Kantianism applied this same idea to human consciousness itself and it is at this point that the neo-Kantian school of philosophy diverted from Kant's original ideas, taking a position that was strongly criticised by Bergson.

For Bergson, the opposition of the ideal (mind) and the real (matter) can be traced to the oppositions between a variety of dualisms: the opposition between the in-extended and the extended, between quality and quantity, between freedom and necessity (*Matter and Memory* p. 244). It is for this reason that Bergson draws a distinction between 'immediate consciousness' (French: 'la conscience immédiate') and 'reflected consciousness' (French: 'la conscience réfléchie'). Our minds function by remembering past states, and anticipating, through imagination, future states. Our minds are thus lodged in the past. They function on the basis of a delay of immediate responses to stimuli. Our physiology, however, is always placed within the immediate present. Physiological mental events are immediate and in the present. They are dominated by impulse and automatism.

Bergson sought to discuss the function of memory not in terms of the brain but in terms of experience and representation: "If pure recollection is already spirit, and if pure perception is still in a sense matter, we ought to be able, by placing ourselves at their meeting place, to throw some light on the reciprocal action of spirit and matter. 'Pure,' that is to say instantaneous, perception is, in fact, only an ideal, an extreme. Every perception fills a certain depth of duration, prolongs the past into the present, and thereby partakes of memory. So that if we take perception in its concrete form, as a synthesis of pure memory and pure perception, that is to say of mind and matter, we compress within its narrowest limits the problem of the union of soul and body" (Bergson 1911 [1907]). In positioning his dualism in terms of duration, Bergson aimed to suspend the human subject between pure spirit and pure matter. He explained this subject as a synthesis of pure material perception and pure spiritual memory. Duration, the ever-evolving and ever-changing nature of things, lies at the basis of both spirit and matter. Spirit and matter are not first order phenomena. Only duration is.

Both materialism and idealism suffer from logical inconsistencies that are related to the inconsistencies of dualism. After all, if we interpret the universe as merely consisting of 'spirit', we end up in a reduction of everything material to our representations thereof. We end up in a representative solipsism. But if we see thinking and sensation merely as an effect of how the physical brain functions, in the sense that the 'dance of the atoms' is somehow translated into psychological contents, human consciousness becomes a side-effect of molecular interactions. Especially the reduction of memory to a function of the brain struck Bergson as a categorical mistake.

An important epistemological convention for scientific research is to make a strict separation between subject and object, between the scientist and that which this scientist observes and experiments upon. This convention becomes problematic, however, when the object of scientific study is the human subject. The conventions within which subject and object are defined come under pressure as soon as the focus of attention shifts towards the study of human consciousness itself. After all, what appears within the scope of our experience is something different that our possibility for experience itself – at least, in most commonly established ontological positions. A key aspect of Bergson's philosophy is what he refers to as immediate data of consciousness. That which is given in an immediate sense is more basic than either the world of phenomena[6] (that which appears to us) or the world of the noumena[7] (that which exists in itself, outside of our experience), but it is even more basic than the subject of knowledge itself[8] (the us to which everything appears). The human mind may capture the intricate mechanism of how a seed grows. Subsequently, the way in which this process is perceived may be studied by studying how the brain registers the shape, scent, colour, etc. of plants. The human soul may be enriched by this experience and the human heart may be touched by the beauty of life, etc., but all these processes presuppose something more primary: the immediacy of experience as such.

In subsequent sections of this chapter, I want to discuss Bergson's view on memory and consciousness in more detail. To do this, I will first of all outline how the philosophical positions discussed above (dualism, deterministic materialism, idealism) influenced scientific thinking, until we reach Bergson's views. Finally, I will discuss how these positions influence debates in contemporary neuroscience. The relevance of Bergson's views on the relation between brain and consciousness lies in the failure of contemporary neuroscience to distinguish consciousness from neural activity. In the next paragraph I will start with an account of the historical predecessors of modern neuroscience: physiognomy, phrenology and other nineteenth century scientific studies of the brain and the nervous system. I will relate these to early materialistic approaches to psychology. In paragraph 5.2 I will critically discusses these on the basis of an elaboration of Bergson's criticism of Kant, specifically of neo-Kantian interpretations of his works. In the last paragraph I will extrapolate the way in which some elements of these nineteenth century approaches persist in some approaches in current neuroscientific research. I will also discuss critics of these approaches within neuroscience. I will do so by extrapolating the relevance of Bergson's critique for contemporary debates in neuroscience. In this

---

[6] That which appears to us, that which is thus experienced within the limits of our sensory perception (also see footnote 25).

[7] That which, in classic philosophy, exists outside of human sensory experience – and in Plato's universe, might be known, but in Kant's universe is indefinitely postponed, behind the horizon of human sensory experience and thus can only be 'thought' (also see Chap. 2).

[8] The self, or the subject's self-experience.

chapter I mainly refer to Bergson's *Matter and Memory* (1896) and *Creative Evolution* (1907), but it will also refer to *Mind Energy* () and *Time and Free Will* (1889).

5.1   The Mechanistic Mind

The Swiss pastor Johann Kaspar Lavater (1741–1801) was one of the first modern scholars to introduce a scientific theory that aimed to relate character traits, thoughts and emotions to physiological characteristics. Similar theories had been introduced in Antiquity, and some attention was given to the topic during the Renaissance, but modern science had, until then, shied away from it. Lavater's approach, 'physiognomics', was a mixture of religion and science. Physiognomics presumed that the divine in human nature could be determined by studying how the mind influences physical features, specifically those of the face. As such, he believed in the possibility to determine the character of a person on the basis of analysis of facial traits. Lavater was a close friend of the writer Johann Wolfgang von Goethe (1749–1832) until the two fell out over the latter's accusing Lavater of superstition and hypocrisy. The idea of a scientific study of the relation between mind and physical characteristics, however, persisted. In the late eighteenth century, the German anatomist Franz Joseph Gall (1758–1828) drafted the basic methods for a research field known as phrenology (Gall 1822–1825). The principle behind phrenology was the assumption that the brain is the organ of the mind. As organ of the mind, it was presumed to consist of independently functioning parts (Yildirim and Sarikcioglu 2007). Gall believed that a person's moral character and intellectual abilities were innate: they were part of how that person's brain was innately organised. The brain, to his mind, was also the organ where our predispositions, emotions and intellectual abilities resided. In his view the different parts of the brain were connected to these different aspects of human functioning: a higher development of one particular talent would, in his view, also be visible in the size of the brain parts responsible for this talent, and these shapes were, to his mind, visible in the external form of the cranium.

Phrenologists presumed they could measure a person's character by measuring the shape of the head. Of specific importance were the lumps and depressions one might find on the skull of the subject involved. During the nineteenth century, phrenology would be further developed by a number of scientists, the most well-known amongst them being Cesare Lombroso (1835–1909). Lombroso was a criminologist, who sought to establish a relationship between specific characteristics of head and face and an innate tendency towards criminal behaviour on the other. He also attached explanatory value to certain facial characteristics. The association between synophrys (having a monobrow) and criminal tendencies stems from his work.[9]

---

[9] Although Lombroso himself distanced himself from his phrenological works later in life, they continued to be influential until well into the twentieth century. His views still hold a cultural

Phrenology stands at the basis of the doctrine of brain localisationism. On the basis of Gall's theories and the earlier scientific research of Julien-Cesar Legallois (1770–1814), the French physiologist Marie Jean Pierre Flourens (1794–1867) made crucial empirical contributions to this doctrine. A pioneer in neuroanatomy, Flourens experiments on rabbits and pigeons demonstrated that different parts of the brain were responsible for different functions such as perception, motoric control and judgement (Flourens 1824; Yildirim and Sarikcioglu 2007). Through lesioning of animal brains, he was also the first to give empirical proof for the location of the mind in the brain, rather than the heart. Until that time, debate still existed over where in the body the human soul and the faculties of the mind could be found. From Flourens onwards, that mind and brain are the same thing became a dominant position. And from here it follows that our talents, our virtues and our vices can be found in different locations of the brain. This also goes for our perceptions, thoughts and emotions.

Brain localisationism remained a core influence in neuroscience for more than a century. Important representatives of localisationism in neuroscience (Moeller et al. 2015) include Paul Pierre Broca (1861) and Carl Wernicke (1874), specifically in their research of language processing, and Théodule Ribot in his work on memory and emotions (Ribot 1881, 1885).

In the late nineteenth century, psychology was maturing into an independent scientific discipline, and different approaches emerged. While Bergson's friend and colleague, the American psychologist William James, was trying to develop an approach to psychology based on introspective empiricism (which he termed 'radical empiricism'), others were labouring to establish a psychology that studied the mind in more analytic and biological terms.[10] Notably Théodule-Armand Ribot (1839–1916) adopted this latter strategy. Ribot was critical of more metaphysical approaches to the human mind. He derided the French spiritualists' attempts to "reach the Absolute through reflection" (Engel 2004, p. 11), opting for an explanation of psychological phenomena through the study of neurophysiology rather than philosophy.

Ribot's *The Maladies of Memory* (1881) treats memory as an aspect of human existence localised in a specific part of the nervous system. He was one of the first to try and bridge mental phenomena and functions such as emotions or memory with the physiology of the brain. Ribot adhered to a materialist thesis, envisaging consciousness to be localised in the brain. Although his thesis is now perceived to be overly schematic, the idea that the mind is the brain is still dominant in contemporary neuropsychology. The study of selfhood in psychology tends to reduce the

---

impact: take for example the monobrow and low forehead of Animal from the Muppet Show, or Lisa Simpson's nemesis, the monobrowed baby from the animated series 'The Simpsons'.

[10] In his approach to pragmatism, James abstains from a distanced and analytic type of empiricism. His empiricism is radical and is beyond the epistemological separation of the subjective and the objective. As Luis Borges phrased it in his discussion of James: "Pragmatism, in the Jamesian sense, does not want to restrict or to lessen the richness of the world; it wants to grow as the world." (J. L. Borges, *Nota preliminar*, p. 12).

self to organised matter, determined by laws of causation. It was this type of reductionism that Bergson considered to be flawed in Ribot's account of memory. It is,
however, a flaw that still dominates discussions over the nature of the mind, specifically in neuroscience.

In spite of the fact that localisationist theories of the mind seem to reflect
Cartesian dualism and its flaws, their influence persists in modern psychology.
Descartes defined matter as 'res extensa', as something existing in three-dimensional
space. He also defined spirit in terms of 'res': a substance. It is here that Bergson
points to a fundamental problem in the implicit metaphysics that also resounded in
the works of Ribot: perceiving of the mind as thing, as substance. The flaw of
Cartesianism should not be sought in its dualism between mind and matter, but
rather in its definition of mind (or spirit) as 'res' (Latin: *res cogitans*), as a substance
similar in nature to matter (Latin: *res extensa*). Descartes double ontology could not
but run into problems. Later ontological positions that reduce spirit (mind) to matter
are dependent of this same substantivist thesis. Bergson's critique is targeted on this
substantivist, and thus spatial, account of the mind.

Bergson, in his response to Ribot outlinestwo types of memory, the one mechanic
and the other physical – the one consisting in a replay or repetition of past actions,
the other serving to retain past events. The latter retains the past in the form of an
image-remembrance. Although this distinction seems to suggest a physical-mental
dualism, it is not the kind of dualism developed by Descartes: it is not a dualism of
two types of substances, the one being material, the other being spiritual.

For Bergson, the mechanical has something comical. Humour is meant to keep
us human and in line with the natural, in spite of the effects of the mechanisation of
the world. It is when humans demonstrate something machine-like, that they become
comical: we tend to laugh when we see a person's gestures as if they were per-
formed by a malfunctioning machine. But the reversal is also true: as soon as the
machine demonstrates human features, it inspires horror: the more a robot has a
human face and the more it mimics human expression, the more uncanny it becomes
to us. The comical is a means to correct the automatisation of our lifeworld. This
also applies to mechanistic views of the psyche: a piano-player who functions like
a pre-programmed machine is either laughable or frightening.

Nineteenth century pseudoscientific approaches such as phrenology and physiognomy took a strongly brain locationalist position towards the faculties of the
mind. Initially, this was also the case for neuroscience, but a shift from locationalism to a more process-oriented conception of the function of the neural system can
be observed in recent years. To be able to sketch out the importance of this paradigm
shift we need to look at the history of neurology in some more detail.

As a modern scientific discipline, neurology was born in the third decade of the
nineteenth century. Important research had contributed to its earliest phases, conducted by scholars such as Gall and later Flourens. The existence of the nervous
system was well known in earlier centuries, and important empirical studies were
conducted by microscopic researchers. But a systematic study of the function of the
brain and the nervous system did not emerge until the 1830s. In 1836, Robert Remak
(1815–1865), described the way in which nervous tissue is embedded in a complex

of so called filamental processes. In the same decade, Jan Evangelista Purkyně (1787–1869), a Czech anatomist and physiologist, described the neural cell. Gabriel Gustav Valentin (1810–1863) and Christian Gottfried Ehrenberg (1795–1876) had already made observations of neurons a few years before Purkyně,[11] but the discovery of the Purkinje cell in 1837 (a neuron with branching dendrites that can be found in the cerebellum) and the Purkinje-fibre (a fibre conducting electrical impulses to animate the heart) remain at the basis of the field of neurology.

Purkyně studied philosophy next to medicine, and was specifically taken with the late 18th and early nineteenth century German idealists. The philosophical interest in this tradition is closely interconnected with his scientific achievements: Purkyně discovered the Purkyně image and the Purkyně shift: the reflection of a perceived object in the eye and the shift in intensity of red and blue at dusk. These discoveries fit well into the criticism of naive realism that is characteristic of idealist philosophy.

Although physioanatomists such as Gall and Flourens contributed much to the earlier studies of the brain, it is on the basis of Purkyně's works that the sciences of the brain truly took shape. Whilst on a superficial glance Purkyně's scientific insights seemed to fit neatly within some of the notions developed by Immanuel Kant, his interest drifted more to the works of German idealist philosophers such as Johann Gottlieb Fichte (1762–1814) and Friedrich von Schelling (1775–1854). His interest in idealism is apparent in his discussion of the phenomenon of synaesthesia, for instance. In spite of his idealist inclinations, his scientific insights also seem to support more dualist traditions in the philosophy of the mind in the sense that perceptions in the brain function as representations of reality: the mind as the mirror of nature. These discoveries made in the early 1800s set the stage for a debate that has still not quieted down. The main reason for its persistence is that the nature of the brain touches on the nature of the mind. In so far as neuropsychology studies the nature of consciousness, it coincides with several problematic issues associated with the dualistic metaphysics of René Descartes, which sees the material and the psychic as completely separate realms. The question how to explain causal relationships between the material and the psychic realm becomes one of the main issues at hand.

Regarded from the perspective of the present state of the art in neuroscience, phrenology is plainly obsolete. It appears as an amalgam of primitive neuroanatomy and implicit moral philosophy. But some presumptions of phrenology still persist. Phrenology assumed for example that specific traits, talents, emotions and functions of the human could be located in specific parts of the brain. It took until the 1980s for a richer perspective to emerge (Moeller et al. 2015), in which the functional role of a part of the brain is not merely given by its neuroanatomical structure alone but rather by its connections to other parts of the brain. Still, localisationism is quite dominant in neuroscience, and in many respects it has been experimentally confirmed. But in some respects this inheritance also carries along certain prejudices that predate the phrenologists of the nineteenth century. This inheritance can be

---

[11] Valentin was the first to describe the cell, nucleus and nucleolus of neurons.

subsumed under the 'mind-brain' dualism, an issue that still puzzles the philosophy of mind.

In *Descartes' Error* (1994), Antonio Damasio provides for a neuro-evolutionary explanation of the faculty of rationality. In his view, capacities that go beyond our instinct, such as rationality, are a result of the increasing complexity of our nervous systems. They are a response to "the daunting task of predicting an uncertain future and planning out actions accordingly" (Damasio 1994). Memory, specifically short-term memory, is a neurological precondition for rationality. Contemporary neuro-scientific research provides for an account of our mind's functioning that does much more justice to the complexity of these processes. Damasio (1994) claims that there is no specific "place" in the brain where different functions such as memory, ratio-nality, sociability, communication etc. can be located. He rather refers to 'conver-gence zones' where signals from different regions in the brain become connected, representing specific internal states and responses to perceptions and environmental triggers. Still, a certain mechanistic reductionism persists for many neuroscientists. It often appears in the guise of the equation of mind and brain. It is here that Bergson's adage still applies: equating mind with brain is a mistake of confusing the metaphysical presuppositions of a scientific field with its conclusions. But drawing out a different basis for neuroscientific research is not an easy task. Have we indeed rid ourselves of Descartes if we replace an atomistic one-to-one view between thought-image and real object by a dualist relation between complex structures in the brain and complex states in the world?

5.2 Memory and Imagination

Although less systematic than other publications (Bor 1990), *Matter and Memory* (Bergson 1988 [1896]) was once regarded as the most important publication Bergson wrote. It influenced a broad variety of thinkers, including Walter Benjamin, Maurice Merleau-Ponty, Jean-Paul Sartre, Emmanuel Levinas, Paul Ricoeur and Gilles Deleuze. Bergson outlines episodic, semantic and procedural memory as dif-ferent types of memory, where others before him had not yet made such distinc-tions. Furthermore, *Matter and Memory* is devoted to resolving a problem that forms one of the most important criticisms of Bergson: the problem that if every-thing should in the first place be conceived of as duration (continuous evolution), one can no longer make distinctions between different entities. As Deleuze states, the past cannot simply be considered as a 'former present'. Nor should one regard the past as a product of the psychological mechanisms of recollection (Deleuze 1966; Ansell Pearson 2004). In other words, consciousness, like time itself, is best described with metaphors that refer to a stream, or a flow, and not with a succession of distinct and discrete states, ideas or perceptions.[12] But if this is the case, how can we then still distinguish between past, present and future, between memory, immediate consciousness and imagination? This problem[13] is specifically poignant in the matter of time: how can past, present and future be distinct if they are part of a constant and continuous evolutionary flow? Ansell Pearson (2014) mentions that William James compared Bergson's revolution to Kant's *Critique of Pure Reason* typifying it as a Copernican revolution as well. Bergson's revolution consisted of considering memory as a specific synthesis of past and present, with a view to the future. But to be able to understand what this synthesis consists of, it is necessary to investigate how Bergson defines the past.

The goal of *Matter and Memory* was to restate the problem of the relation between thing and thought, and thus come to an alternative to the way this problem was treated by Immanuel Kant. In this treatment, the notion of immediate experience is of central importance to Bergson's objections to this Kantian treatment. But to be able to do so, Bergson needed to give an account of memory as well. Immediacy is not yet memory. It is pure perception, which is absolutely individual in nature. This individuality of pure perceptions can also be seen as their "thisness" or, to use a term coined by Duns Scotus (1266–1308), their haecceity. But where Duns Scotus considers haeccaeity as an intermediate between the real and the conceptual, Bergson considers it as the intermediate between actual and virtual (Image 5.1).

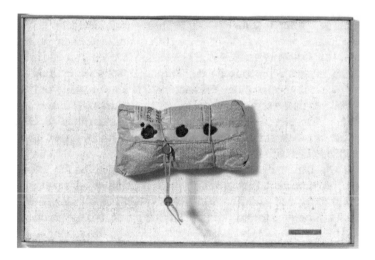

Image. 5.1 Piero Manzoni (1933–1963), *Achrome*, 1961, pacco in carta di giornale, 40 × 60 cm

---

[12] It is from this issue that Bergson's negative assessment of cinema derives – a view that was challenged by Deleuze.

[13] Which, in a different guise, can also be found in the works of Derrida (the issue of the activity of differing, as différance, with an 'a', rather than the passivity of different states (as difference, with an 'e'): see Chap. 7.

Following a somewhat superficial reading of Kant, one might be tempted to conclude that pure sensory experience is without any form or shape, and that our recognition of such experience is perception. This would then consist of looking at a series of holiday slides from the Spanish Costa Brava. Memory would be the looking at these slides in some kind of chamber of consciousness. But for Bergson, memory should not be considered as "an observation of pictures" (Tymieniecka 2009, p. 237). A more conscientious reading of Kant's approach to the matter of memory would run along the following lines: recognising something *as* something is a necessary precondition for representation. This necessitates memory: to be able to relate an observation in the present to an observation in the past, and thus constitute something as an object, of a certain number (e.g. one), with specific quantity, quality, modality and relationality (see Kant, *Kritik der reinen Vernunft A*[14] (*Critic of Pure Reason A)*, pp. 104–105). After this account of recognition, in its formal nature, Kant discusses his famous conception of apperception: this is the conception by which Kant manages to show how we can come to cognitive judgements (over our recognised perceptions) at all. It supports the idea that we need to perceive our perceptions to be able to come to such judgement. To phrase it differently: perceptions of our perceptions can only occur if we follow a certain rule, namely that I acknowledge that I am the one to have such perceptions. It should always be the case that an 'I think that' guides our judgements. In a sense, this is Kant's formal way of expressing that if we want to come to certain judgements, we need to recognise an object *as* an object, and ourselves, the subject, as the subject that makes that judgement. Thus, Kant was relieved of the necessity of showing how the mind adjusts to objects (a necessary consequence of Cartesian epistemology), and instead objects could be posited as a function of knowledge (a turning around of metaphysics that he refers to as a Copernican turn). This latter part sums up Kant's transcendental argument for apperception: it is here that unity and identity of the knowing subject is tied into its consciousness of itself as the subject of all its representations: only from this point can one justify the existence of judgement at all. Objects then do not appear as pre-existing knowledge autonomously: they appear as a function for knowledge in this edifice instead.

Bergson was extremely critical of the Kantian edifice for the justification of knowledge. He perceives a Parmenidean[15] undercurrent in Kant's metaphysics. His main point of difference with Kant's in Kant's system, sensory experience as such can never be valid source for knowledge. But to Bergson, this means that all knowledge would be constructed knowledge, even to the extent that space and time are, in Kant's view, pure forms of intuition, so that perceiving things in space and time remains a construction on the side of the subject.[16] The objection made by Bergson

---

[14] The 1781 version is usually referred to as *Kritik der reinen Vernunft A*. The section referred to here can only be found in this first version of the *Critique of Pure Reason*.

[15] As stated in the introduction of this book, Parmenides was the Greek pre-Socratic philosopher who posited that all is fixed and no change or differentiation ultimately exists.

[16] It is at this point that it becomes questionable to what extent Bergson criticises Kant himself, or Kantian scholars' interpretations of later generations whilst thinking he criticised Kant.

is that, due to Kant's view, an unjustified separation is made between phenomenon and noumenon, subject and object, between mind, body and the world (Ansell Pearson 2002, p. 12). It is unjustified since it can only be made on an abstract level, in formally stated conventions and conceptions. And without concreteness it is of no value. But trying to render it concrete leads to even larger problem, as we have seen in the above discussion of neo-Kantianism, because time is here considered as a kind of time-line (a spatial construct). Bergson considers time – duration – as fundamentally different from space. He does not perceive of it as a pseudo-geometrical category but as a convergence of duration, life and intuition.

If Kantian epistemology starts from analysis and differentiation of forms of intuition, categories of recognition etc., Bergson starts from pure and immediate experience, since formal possibilities for knowledge already necessitate an account of experience as mediated. It is here that the basis of Bergson's views on the nature of memory can be situated as well. Whilst perception turns a mere image into a representation, our representation of perception in the form of memories has a duration itself. It is an unconscious process. It is memory that makes perception conscious.

Bergson distinguishes between habitual memory (*mémoire habitude*) and pure memory (*mémoire pure*). Habitual memory is related to the sensory-motoric system and enables us to walk, sit down and pick up a glass. It is also the type of memory that enables us to recite a verse or know our way through our home town. Pure memory, in contrast, is enabled by an imaginative perception. It belongs to the virtual rather than the actual world (Bergson 1988 [1896], p. 82). To return to the Costa Brava, Bergson draws a parallel between pure memory and image remembrance: it recognises the past as past, and it is in this respect that it also recognises the past as unchangeable: the holiday in Spain was a special experience, but it cannot be altered or revisited (except via a new holiday of course). It is dynamic, constant, spontaneous and, most importantly, it is representational in nature. It does not start from conditioned behavioural patterns but it represents the world.

For an account of free will, reference to memory alone is not sufficient as a philosophical ground. To be able to place ourselves, our actions and our plans in relation to our memory alone does not yet reveal how we are able to act freely towards a future. Humankind has the faculty of memory as well as the faculty of imagination. This frees us from the confines of the natural world, be it those of the inorganic or the organic. We can both remember past situations and imagine possible futures, and we may align our actions according to specific preferences and insights derived from this. Imagination is the faculty that enables us to separate reality – perception, lived, sensory experience – into images. Sensory awareness is only given in its immediacy in the actual present. Sensory perception is not some type of consciousness-internal representation of some external phenomenon (Bor 1990, p. 81). Our immediate sensory awareness is not (yet) an 'awareness of', for this would imply again a separation between things and mind. At this stage no boundaries between 'subject' and 'object' are drawn yet: these are constituted in reflected experience, they are not yet present in immediate, concrete awareness. The subject/object division only follows later.

Bergson's *Time and Free Will* (2001 [1889]) focuses on the nature of free will and its intertwinement with his notion of time as duration. His views on the matter remain constant throughout his works, and also apply to human experience. The human mind should not be regarded in spatial terms, as some kind of chamber of consciousness. It exists in relation to time rather than to space. And the reduction of the human mind to the category of space is the key reason why we cannot accommodate a concept of free will in both past and current studies of the brain. In such reductions, we again reduce the temporal aspect of human being to the artificially objectified notion of time as it exists in physics: clock-time (see Chap. 2). It is this view of time which also functions in psychological experiments. However, for a proper characterisation of the human mind and human existence – and here Bergson's ideas run quite parallel to those of Martin Heidegger – we need to look into the different ways in which our existence is shaped by and typified by time. For Bergson, this means that we need to acknowledge memory and imagination as intertwined modes of – again, anachronistically put in Heideggerian terms[17] – our being in the world. For Heidegger, individual authenticity is a result of our confrontation with our existential 'Angst' (mostly translated as dread or anxiety) in which we realise the uncanny dimensions of existence while at the same time relating to ourselves from within our being, taking our being-in-the-world as point of departure. For Bergson, this focus on the individual self in his or her 'thrownness' into the world is not the most fundamental aspect of our being. Instead, Bergson sees the experience of duration as the basis for a metaphysics, rather than the experience of world and self in a self-referential sense.

For Bergson, neither traditional teleology nor material determinism can provide for an acceptable account of free will, and, as such, of human agency. Teleology takes the view that the entirety of the universe, including evolutionary processes, is the actualisation of some kind of predetermined plan. Material determinism also holds that the way in which reality unfolds is pre-determined, but not according to some kind of plan. Bergson accepts neither, since both depart from a presumption of the nature of time as similar to space. Bergson's notion of freedom is the relation of the 'concrete self' to 'the act which it performs'. To his mind, this relation cannot be seen as pre-determined, because this entails again a determinism based on the notion of a time-line: a causational view of time. In other words, freedom is a kind of creative process. Determinism presupposes two events on a given time-line: a determining and a determined event. The experience of time as duration, however, does not assume such linearity and involves unpredictability. In the case of determinism, we replace process experiences by notions derived from extension, therefore ignoring the true nature of what we study.

Materialist determinism perceives of everything, including consciousness, as determined by the workings of atoms, molecules and, eventually, the basic laws of physics. Bergson criticises this position since it applies a concept of time to living

---

[17] Martin Heidegger was rather critical of Bergson's work, but his reading of Bergson's oeuvre is rather unfair and deficient at many points (see Massey 2015).

systems which does not do justice to their process-characteristics. To be able to articulate this criticism, Bergson saw a need to define a vocabulary of his own. The material determinist for example would not refrain from using concepts like 'mental states' in discussing the nature of consciousness. For Bergson, such a concept is problematic since it freezes the present, as if time would consist of instances considered as dots on a line, a sequence of states. In creative evolution, there are no fixed states.

Bergson's criticism of material determinism, or mechanistic thought, is that it considers time as something fixed. It cannot account for the dynamics involved in evolutionary processes. As Bergson wrote in *Time and Free Will*; for the mechanistic thinker "[…] any principle is simple of which the effects can be foreseen and even calculated: thus, by the very definition, the notion of inertia becomes simpler than that of freedom, the homogeneous simpler than the heterogeneous, the abstract simpler than the concrete." (Bergson 2001 [1889], p. 141). Bergson further states: "As […] the principle of the conservation of energy has been assumed to admit of no exception, there is not an atom, either in the nervous system or in the whole of the universe, whose position is not determined by the sum of the mechanical actions which the other atoms exert upon it. And the mathematician who knew the position of the molecules or atoms of a human organism at a given moment, as well as the position and motion of all the atoms in the universe capable of influencing it, could calculate with unfailing certainty the past, present and future actions of the person to whom this organism belongs, just as one predicts an astronomical phenomenon." (Bergson 2001 [1889] p. 141). This view is still the position of many neuroscientists today. But time in the sense of concrete time cannot be broken up into different instances, since this already would reduce duration to spatial categories. And abstract time, the concept of time that is suggested in most theories in modern physics, and materialises in our time-measuring devices – clocks – does not apply to lived experience. It merely applies to the theoretical concepts of classical mechanics and to our own pragmatic need for time measuring devices.

Bergson still adheres to a kind of physical-psychic dualism. Throughout his life, he becomes increasingly nuanced on this position. As we have seen, Bergson's dualism does not consider the spirit as a substance although duration still is. Substance however, should here not be interpreted as 'what lies underneath' nor as 'thing', but rather as 'that which cannot be further reduced'. Our consciousness is not spatial but temporal in nature: we should thus not follow a dualism between two types of substances (mind and matter) but a dualism between spatial matter and enduring time. But our sensory experience and the ensuing perceptions are contained in a material world, and this appears to create a paradox. The oscillation between perception and memories should therefore be understood in its proper nature: where perceptions carry along a spatial dimension, memory entails a temporal dimension. And if one regards memory as specifically temporal in nature – as being related to enduring processes – it is no longer necessary to conceive it in spatial terms or as having a specific location somewhere inside our skull.

The subject-object division is functional for scientific thinking because it allows to formalise positions in the context of an experiment. At the same time, it is responsible for the epistemological tendency to develop a dualist metaphysics. The acceptance of a dualist worldview – distinguishing between the mind's free will and the causally determined world of things – stems from this division. The next epistemological step, and the most fatal one for Bergson, is then to define subject and object in terms of two 'substances'. This is the main deficit of the Cartesian heritage. It is also responsible for the very idea of determinism. Subject and object can only emerge after immediate experience. This also goes for determinism and free will. These binary concepts are not immediately given, but reflect an implicit metaphysics.

Bergson is critical of the metaphysical positions of idealism and materialism (in his phrasing: realism) as suffering from the same implicit dualism. They tend to contrast perception and conceptualisation, whilst implicitly claiming an absolute truth validity position without giving the proper arguments for that validity. For Bergson, the problem of the parallelism between mind and brain is a philosophical illusion (Bergson). Mental contents do not have parallels in brain structures. The reason we often cast our views on our minds in such phrasings is still due to Cartesian dualism, a position that in spite of its popularity as a 'common sense' philosophy is riddled with paradoxes and logical inconsistencies. Bergson captures these contradictions in *Matter and Memory* as follows: "[T]o state this proposition is enough to show its absurdity. The brain is part of the material world; the material world is not part of the brain." (Bergson 1988 [1896], p. 4).

Quite often, we either define an idealist or a realist answer to the relation between mind and brain. This means that either everything belongs to the ideal world (so that talking about a brain merely means talking in terms and concepts generated by neurophysiological discourse), or everything belongs to a material world of neurochemical patterns (including the claims generated by neuroscience). Bergson, however, reasons that we first need to accept that any description of reality is either idealist or realist. We also need to accept that both positions are mutually exclusive. Subsequently, we need to accept that any idealist position embracing a mind-brain parallelism is self-contradictory if it does not ultimately takes a realist point of view, thus losing its idealist position. But similarly, a realist position embracing a mind-brain parallelism is self-contradictory if it does not ultimately take an idealist position, and so on.

Parallelism is not logically defensible. Still, many approaches in neuroscience take precisely this parallelist position, usually departing from a realist-materialist assumption, considering mental phenomena as 'epiphenomena' (phenomena emerging from more fundamental physical phenomena). As Bergson writes in *Time and Free Will*: "[…] the most radical of mechanical theories is that which makes consciousness an epiphenomenon which, in given circumstances, may supervene on certain molecular movements. But, if molecular movement can create sensation out of a zero of consciousness [out of no consciousness at all], why should not consciousness in its turn create movement either out of a zero of kinetic and potential energy, or by making use of this energy in its own way?" (Bergson 2001 [1889], p. 152). This epiphenomenalist view is an attempt to resolve the problem by holding that mental events such as thoughts, consciousness, emotions, reflection etc. are fully caused by physical events (the function of the senses, the intricate workings of our brain parts or other organs, our neural impulses, and so on). As such, it holds that mental events are mere illusions. They are neither cause of other events, nor part of some ideal world. Epiphenomenalism, as a form of materialism, suffers from the same problems as any form of dualist parallelism, in spite of its denial of the existence of mental phenomena. It also intermingles the incompatible metaphysical frameworks of idealism and realism.

According to Bergson, to explain what 'the mind' is, as we have seen in the preceding paragraphs, we cannot retreat into the monism of materialistic realism; nor can we retreat into a pure transcendental idealism. A simple dichotomy between the mental and the cerebral is not tenable either, due to its dualist presumption of two substances. Related accounts that try to retain a materialist-realism are forced to posit the mind as an epiphenomenon to account for the experience we have of our mind. But, as Bergson showed, this position suffers from the same problem as the original non-epiphenomenalist approaches to materialism. An added problem is the consequence for neuroscientific knowledge itself: if one regards the mind as merely an epiphenomenon, the same goes for the contents of the mind: emotions, associations, but also thoughts, ideas, theories. Thus, the contents of neuroscientific theories themselves become nothing more than an 'epi-epiphenomenon' in their own right. They are merely an accidental side effect of an accidental side effect. Descartes got stranded when he, in Bergson's words (*Matter and Memory*) "set up the mathematical relations between phenomena as their very essence", thus being "obliged to regard the mathematical order of the universe as a mere accident" (Bergson 1988 [1896], p. 82). Similarly, this epi-epiphenomenological position inadvertently declares its own theoretical position invalid.[18]

The Cartesian view of the mind, as we have seen, is riddled with problems and paradoxes. Still, Bergson does make a problematic distinction himself between action on the one hand, which is supposed to be located in the brain, and memory on the other, which he conceived of as spiritual in nature (Petit 1997). It remains clear however that Bergson opposes the view that memory resides in a specific compartment of the brain. Our notions of past, present and future are part of a continuum. Our distinctions between them result from different types of acting memories. As Başar and Düzgün (2016) also discuss, this implies that memory and mind cannot be distinguished. This appears to be confirmed by neuroscientific research of the past decades: Fuster (1995) also conceives of all brain functions as memory, and Bergson reported that memory is the mind itself. Our orientation in the world consists of past memories and present memories as well as imagination: these are needed to consider possible futures. This might seem to suggest that time itself is construed by the function of memory. In a sense this is true, but only to the extent to which we have this faculty in our orientation within the world. Absolute time is, for Bergson, as discussed in chapter three, an abstract construction, based on the

---

[18] The problem referred to in this chapter as 'epi-epiphenomenalism' is the same problem on the basis of which Edmund Husserl criticised several aspects of the neo-Kantian school in philosophy (see 4.1).

automatism to regard all experience in spatial, geometric terms. Concrete time, however, manifests itself in change. Here we arrive at an important issue: the abstracted mechanistic view on time is a construction, determining the course of the history of the exact sciences, but resulting from a very specific inbuilt orientation on past, present and future. Starting from a frozen view of time and of the past, material determinism subsequently extrapolates this same view of time upon the future. If we see the past in terms of creative processes, however, the faculty of imagination, as a basis for anticipation, becomes a precondition of free will.

Finally, it might be useful to draw a parallel to Taoist philosophy here. In this oriental tradition, nature is change and change is nature. Thus, any notion of free will or determinism becomes meaningless. Free will should therefore not be understood in its meaning of absolute free choice, or absolute autonomy, but rather in terms of steering one's course over the naturally emerging paths through the woods. We are part of the world. As Einstein, paraphrasing Arthur Schopenhauer's *On Free Will* once stated: "Man can do what he wills, but he can't will what he wills"[19] (Einstein??). Our will is not a violation of the natural order, but rather an intrinsic part of nature as a creative process. When we no longer accept the conventional division between subject and object, we no longer need binary notions such as free will or determinism. They only emerge as a result of setting such binary conventions.

5.3 Unlocking the Chamber of Consciousness

Views such as those expressed in the works of Théodule Ribot discussed above for a long time dominated the studies of the brain, but a shift in perspective occurred roughly from the early 1990s onwards. Cognitive neuroscience, neuropsychology and neurophysiology no longer seek a correlation between specific parts of the brain and specific concepts, emotions, thoughts, ideas or perceptions (Colbrook 2014). There is no one-to-one relation between objects in the world and images in the brain. The brain rather functions in terms of complex relations.

In "Note on the situation of biological philosophy in France" (1947), the French philosopher Georges Canguilhem (1904–1995) stresses the value of Bergsonian philosophy because it aims: "to understand the true relationship of organism and mechanism, to develop a biological philosophy of machinism, to conceive machines as the organs of life, and to lay down the base of a general organology." (quoted from Wong 2014). In this sense, machines should be understood, according to Canguilhem, as an exponent of our organic systems. Canguilhem terms this Bergsonian approach a general organology, on the basis of the following fragment:

---

[19] Einstein, in *The World as I See It* (1934) used this paraphrase as a quote from Schopenhauer's *On the Freedom of the Will* (1960 [1839]). But the phrase "Der Mensch kann tun was er will; er kann aber nicht wollen was er will" cannot be traced in Schopenhauer's essay.

"If our organs [our limbs, our intestines, our brain] are natural instruments, then our instruments must be considered as artificial organs. Machines which run on oil or coal have imparted to our organism an extension so vast, have endowed it with a power so mighty, so out of proportion to the size and strength of that organism, that surely none of all this was foreseen in this structural plan of our species. Now, in this body, distended out of all proportion, the soul remains what it was, too small to fill it, too weak to guide it. Hence the gap between the two. Hence the tremendous social, political and international problems which are just so many definitions of this gap, and which provoke so many chaotic and ineffectual efforts to fill it. What we need are new reserves of potential energy – moral energy this time" (Bergson 1977 [1932]).

One point of departure for an alternative is to stop perceiving the mind as something related to passive consciousness, and instead see the mind as a mind in action. Brain states cannot be equated or paralleled to perceptions and perceptions cannot be equated or paralleled with states of the world. Instead of parallelism, interaction. Nonetheless, some three decades years earlier, Bergson wrote: "As soon as we compare the structure of the spinal cord with that of the brain, we are bound to infer that there is merely a difference of complication, and not a difference in kind, between the functions of the brain and the reflex activity of the medullary system." (Bergson 1988 [1896] p. 10). But both these systems, for Bergson, are not 'things' or substances, but processes. In terms of the context of this remark in the debate on determinism and free will, Bergson points out that he does not see how sensory input might lead to representations in the mind in a causal fashion (this also being the criticism which Kant juxtaposed against Humean views on perception[20]). He does however see how the cells involved in the regions in the cortex associated with perception may reach the motor system 'at will', thus 'choosing' the effect sensory input might have. Thus, Bergson embraces a view on the brain as an instrument for action, not as the – materially determined – locus of consciousness itself. The brain thus should not be regarded as identical to the mind.

In the third chapter of *Event* (2014), the Slovenian philosopher Slavoj Žižek discusses how the reductionist view on the mind (the mind as the brain) remains an attractive position. Referring to Jürgen Habermas, he alludes to the problem that although neuroscientists appear to have uncovered the neurobiological 'proof' against free will, free will is in all actuality a necessary precondition of the scientific mind, of rational agency. This means that reductionist approaches to the mind that interpret mental events such as thought, memory and imagination as mechanically determined functions are in conflict with the precondition of taking a rational approach to these phenomena as such. In spite of this fundamental inconsistency, many reductionist neuroscientists still prefer mechanistic approaches when it comes to analysing the functioning of the brain. This internal epistemological conflict arises due to the omission of a notion of duration – thoughts, memory and imagination are reframed as 'brain states'. They are isolated and disconnected from the way they evolve over time. To be able to develop an alternative, the function of memory

---

[20] David Hume posed a causal link from object in reality to impression on sense organs to representation of reality in the mind.

and imagination need to be researched from within, rather than as an external material substance. Bergson adheres to a non-materialist view on memory: memories should be understood in terms of their relation to time (as duration, as a creative process), not in terms of their relation to the grey gooey mass found in our skulls.

The implications of neuroscientific discoveries have not sufficiently been integrated in philosophical explanations of brain and mind, self-identity and free will. Neural man, as Catherine Malabou, phrases it, still has no consciousness. In spite of progress in the state of the art in the neurosciences, the image that these sciences have created of humankind's neural functioning lacks an explanation of what consciousness amounts to: it cannot grasp this dimension of human identity and human functioning, and this although we should conceive of consciousness as the most recognisable aspect of our self-identity and our functioning. As we have seen, this leads some to ignore the issue, reducing consciousness to an epiphenomenon.[21] But this strategy is not satisfactory, philosophically speaking. As Malabou states: "We are still foreign to ourselves [...]. 'We' have no idea who 'we' are, no idea what is inside 'us'" (Malabou 2008, p. 3). Explanatory frameworks still adhere to a rather static perspective on the functioning of the brain. As Malabou puts it: "Our brain is plastic, and we do not know it" (Malabou 2008, p. 4).

To some extent we still like to think of our consciousness as a chamber that can be localised somewhere inside our brain – with us presiding over it in a helmsman's seat. The chamber is a room with a view: it holds two windows faced outwards. It is furnished with two speakers on the sides that carry in the sounds from the outside world, picked up by two ear-shaped microphones, and holds some type of control panel by which we can direct the movements of our limbs, hands, feet, fingers, toes and other exterior organs. We imagine a screen whereupon our perceptions emerge upon, with us as the audience witnessing the shadowy puppets portrayed on the white canvas. Bergson was well aware of this tempting metaphor and identified it with our tendency to think conscious awareness in terms of space rather than duration. Specifically, the then current accounts of psychophysiology remain trapped within the spatial presumptions concerning conscious awareness.

The contemporary neurosciences embrace an interdisciplinary approach, combining psychology, biophysics and neurology. But they still seek a comprehensive position in which all of these can be based on a neurophysiological fundament. In this view, neuroscience essentially remains reducible to neuro-biochemistry. Therefore, this idea of a comprehensive or even holistic approach in the neurosciences merely remains, to quote Marcus Jacobson's *Foundations of Neuroscience* "on probation until such time as a complete reductionist program can be mounted" (Jacobson 1993, p. 141). In terms of contemporary debates on neuroscience, this belief in an ultimate reductionist program for neuroscience remains dominant (although not without challenge, e.g. O'Regan 2011). The materialist thesis still is the essential presumption of many research lines in neuroscience. An example can

---

[21] E.g. neo-Kantians like von Helmholtz, or more contemporary exponents of such causal materialist positions like Daniel Dennet or the Dutch neuroscientist Dick Swaab.

be found in the European 'Human Brain Project' that was launched in 2013, initi-ated by neuroscientist Henry Makram to attempt a simulation of a functional brain. The human brain project is criticised for its attempt to simulate the entire human brain in a computer. Some colleagues state that it is too premature to attempt such a simulation, given the current scientific state of the art. Others, however, see a more fundamental problem in the presumption that mental states and neurological sys-tems can be equated and translated in silico, as information structures in a computer. One way to address the conceptual impasse in which neuroscience finds itself today might be to reconsider the temporal aspects of human consciousness (Image 5.2).

We are still accustomed to the idea that memories are somehow 'stored' in our brains, as a material object, or at least, an information superstructure that resides on this material basis. But rather than perceiving the brain as a container of thoughts, memories, emotions and perceptions, it should rather be seen as its mediator (Middleton and Brown 2005, p. 78). Mechanistic neuroscientific accounts of mem-ory have overly considered memory as – in Bergson's critical wording – 'a cerebral deposit' of past experiences. Memories are then stored, similar to data in computers. Although this can be investigated empirically by looking at altered responses and altered reactivity, phrasing memory in terms of data storage remains metaphorical. The assumption that material causes act on perception and that perceptions are then somehow stored in the brain, is untenable. Furthermore, the presumption of some

Image 5.2 Marcel Duchamps (1912) Nu en descendant un escalier

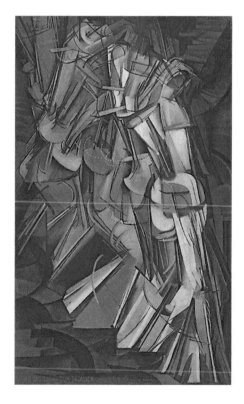

kind of 'storage' has never been empirically clarified. It remains a hypothesis at best. As a result, brain research is often presented as a final solution to age-old philosophical dilemmas, in a sense replacing philosophy, by elucidating topics such as awareness, conscience, consciousness, intentionality, free will, morality, empathy, affection etc.. Neuroscientific research extends the boundaries of the exact sciences, gradually colonising topics and themes traditionally thought to reside with the humanities. Neuroscience thus seems to enable a resolution of issues related to free will, the nature of human consciousness, the relation between mind and body and even the virtues and vices of human beings. But the brain is not identical with the mind, it is merely the instrument through which the mind can manifest itself.

Given the fact that evolution has shaped our brains, there is tendency now to explain anything related to human experience on the basis of evolutionary advantages. Here again, we encounter an issue of an epistemological rather than of a scientific nature. What does 'explaining' mean in this regard? If we are able to provide an account of a given experience or action (say, the emotion felt in seeing an artwork in a gallery, or the interior state of the mind during a scientific Eureka-moment) in terms of neurosynaptic responses or phenomena in brain activity that can be represented through neuroimaging, have we then said anything about the artistic value of that work of art, or the validity of the insight intuited by that scientist in question? No. It rather appears that one paradigm – that of neuroscience – is used to give an account of phenomena belonging to another paradigm (theory of art, scientific validity etc.) without accounting for the fundamental distinctions between these paradigms.

In theories adhering to rationality, emotions are often considered to be a hurdle for rational agency. In *The Two Sources of Morality and Religion*, Bergson discusses the role of emotions in intuition. He sets out with an account of the relationships that exist between emotion and representation. Normally, the process begins with some kind of representation that triggers the experience of a particular emotion. Thus, to use Bergson's example, our representation of a friend might be the cause of an emotion of joy. But Bergson also posits a 'creative emotion', in which we first have an emotion which then generates or at least colours a representation. This is what Bergson considers to be at the basis of the arts. A composer, for example, may first of all experience a particular mood or emotion, and then elaborate it into a composition based on that emotion.

Only from a materialist paradigm can it appear as acceptable to posit a one-directional causal chain from external phenomenon to impulse, to emotion to representation. But such a causal explanation of the emergence of mental representations is, as we have seen in the preceding chapters, a result from the implicit adoption of a specific metaphysical predisposition. It is not a scientific outcome. Neuro-reductionism is such an epistemological choice for a specific metaphysical predisposition. It does not follow from natural observation that we are 'mere' enzymes and neuro-synapses: such a contention follows from an a prior choice in favour of a specific position, which then serves as a framework for interpretation. This is not necessarily wrong. The choice for a reductionist model may aid, for instance, the identification of specific dysfunctions in the brain, and thus contribute to the

Image 5.3 Electro-
microscopic picture of
neurosynapses
communicating

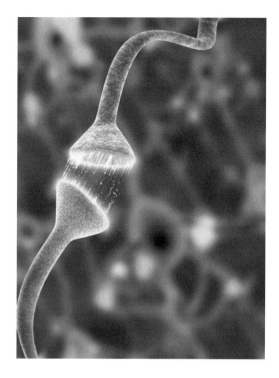

detection or alleviation of a pathology. But one should remain aware that implicit metaphysical positioning comes at a price: it informs an ontologically invalid theory of human identity. This invalidity is specifically poignant in the materialist paradox: the conviction that materialistic knowledge contents are in themselves nothing more than mechanically determined epiphenomena produced by the contingent structure of our brains.

In *Time and Free Will* (2001 [1889]) Bergson makes a distinction between 'time flown' and 'time flowing'. He contends that Time that has passed is determined and can be represented with the vocabulary of space, whilst time in its current flow is undetermined, and cannot be represented in terms of spatial concepts: it is evolving and as such categorically different from space. In this vein, it is not predetermined either. As Bergson himself wrote: "[…] every demand for explanation with regard to freedom comes back, without our suspecting it, to the following question: "Can time be adequately represented by space?" To which we answer: Yes, if you are dealing with time flown; No, if you speak of time flowing. Now, the free act takes place in time which is flowing and not in time which has already flown. Freedom is therefore a fact, and among the facts which we observe there is none clearer. All the difficulties of the problem, and the problem itself, arise from the desire to endow duration with the same attributes as extensity, to interpret a succession by a simultaneity, and to express the idea of freedom in a language into which it is obviously untranslatable" (Bergson 2001 [1889], p.221). In other words, human freedom disappears as soon as the categories traditionally attributed to space, and used to explain space, are applied to time.

Memory and imagination constitute the basis of our ability to act freely. At the same time, metaphors generated by our imagination are responsible for propagating wrongful conceptions of time. Our projection of a fixed and frozen past leading via a pre-determined present towards a future creates an image of time as if it were a line – the present being a dot moving from past to future. But time is first and foremost lived time. Time cannot be isolated from the complexity of evolving structures in which it becomes manifest. Given the fact that Bergson considers clock time as merely a metaphorically reduction that enables a projection of time in terms of space, a deeper understanding of time must be key to understand the interrelation between these different forms of life, up to human freedom.

The issue of free will is one of the traditional quandaries of philosophy. It is interrelated with moral responsibility as well as with concepts of sin and guilt. Conceptually speaking, it does not only conflict with determinism, but also with randomness, with order and chaos. Even before any documented philosophical ponderings over human nature emerged, the notion of free will and that of determinism were already of key importance to human self-understanding. The myth of Oedipus clearly reflects the struggle between fate on the one hand and human free agency on the other. In the case of Oedipus, his murder of his father and his marriage to his mother was realised in spite of (or rather, because of) his very attempt to flee his fate. No sense of moral agency can exist without the assumption of choice. The dilemma between determinism and free will lies at the basis of modern scientific thinking. Although the interpretation of phenomena in the world as being causally determined gave rise to modern scientific thought, the idea of scientific discovery cannot be upheld without a presupposition of free rational agency at the side of the scientific agent himself.

The dilemma between determinism and free will is, for Bergson, the result of a theoretical relation drawn from the 'unextended' – the qualitative – into the 'extended' – the quantitative (Bergson 2001 [1889] p. xiii and 70). With this wording, Bergson points towards the Cartesian frame of mind, in which world and experience are explained on the basis of a terminology derived from three-dimensionality, with a disregard of the temporal nature of experience, change and growth. We take human consciousness – something intricately bound up with perception, memory and time (duration) – and interpret it in terms of extension: as something that can be located in three-dimensional space. Subsequently, this is extrapolated into the idea of consciousness as something which exists inside the bony dome of our cranium. But it is irrelevant that the size of the human brain is much larger than that both that of a spider[22] whose intelligence is caught in a case no larger than a grain of salt or that of tyrannosaurus rex, whose intelligence was caught in a case no larger than a walnut. Thought occurs in the bowels and the heart as much as it does in the brain.

---

[22] Some demonstrate highly complex behaviour: jumping spiders perform complex ritual mating dances; they mimic prey to lure other species of spiders out of their hidings; they have hunting tactics superior to a cat.

It not size or location that forms the precondition of that self-awareness and world-awareness that are uniquely human. The experience of freedom can only be adequately explained in terms of *duration* and not in terms of *location*. It is supposed to exist in animal life lesser so, in plant life hardly so, and in inorganic elements the least. It emerges out of our imaginative remembering of past events and our focus of action towards the future. As a result, both free will and determinism are only meaningful within our discursive relationship with the world, and not as a localisable feature of our brain.

If we focus our attention on the experiences of human beings, we do not leave the worlds of planets, crystals, plants and amoebae behind, but this does not mean that human consciousness can be wholly explained in terms of matter and extension. This materialistic tendency, implicit in behaviourist psychology, neuropsychology and more traditional approaches in cognitive neuroscience, inevitably ends in self-contradiction, seeing the mind as an epiphenomenon caused by biochemically determined brain processes. And here, life (concrete experience) and scientific knowledge seem to represent incompatible perspectives. But this only holds if we consider 'knowledge' to be restricted to and identical with 'rational knowledge', i.e. reductionism. Yet, as was indicated earlier, there are two ways of knowing an entity: by going around it, studying it from all angles, or by entering into it, studying it from within (Bergson 1903). The first type of knowledge is analytic knowledge concerning various aspects and components of the object in question, which fails to truly grasp the object in its own individual nature. The second type of knowledge is intuitive in kind. It can, however, only fully exist in our own individual experience. The entity we ourselves are, is the only thing to which we have a self-evident immediate access. But this does not mean that intuition has no role to play in other types of knowledge. Specifically with regard to our functioning as humans, the analytical way of knowing should always be tempered by our intuitive awareness of what we are. In this regard, the study of the filaments, the bowels and the nerves can never fully reveal our immediate experience.

Humans are characterised by a distinctive creative capacity for self-knowledge and self-adjustment. Intelligence is not a human privilege, however, but something that evolves progressively. The mind or spirit does not host abstract, timeless and universal ideas and concepts. As such, intelligence is lived rather than thought. In recent years, different views on the functioning of the brain have emerged. From the 1990s, several important developments have taken place in neuroscience. As Claire Colbrooke remarks: "Contemporary cognitive science and certain philosophies of the human have drawn upon anti-Cartesianism to insist that man is not a camera, not a computer, and the eye is not a window." (Colbrook 2014 p. 14). Drawing on molecular biology, evolution biology, artificial life, complex systems theory, neuroscience and psychology, Evan Thompson (2007) discusses not the binary opposites of mind and brain, but rather those of mind and life. Here the brain is, as the mind, part of a more complex temporal structure. This opens up new venues for research that do not necessitate a mechanistic approach to human functioning or a reduction of mind to matter.

Literature

Ansell Pearson, K. (2004). *The curious time of memory*. Paper presented to the 7th International Conference of Philosophy, Psychology, Psychiatry. University of Heidelberg, September 2004.

Ansell-Pearson, K. (2018). *Bergson: Thinking beyond the human condition*. New York: Bloomsbury Academic.

Başar, E., & Düzgün, A. (2016). The brain as a working syncytium and memory as a continuum in a hyper timespace: Oscillations lead to a new model. *International Journal of Psychophysiology, Volume, 103*(May 2016), 199–214.

Bergson, H. (1908 [1889]). *Essai sur les données immédiates de la conscience*. 6th ed., 1908, WS, IA. Doctoral dissertation.

Bergson, H. (1988 [1896]). *Matter and memory*. Translated bys N. M. Paul & W. S. Palmer. New York: Zone Books.

Bergson, H. (1912 [1903]). *An introduction to metaphysics* (trans; Hulme, T.E.). Hackett Publishing Company 1999. Also included in: Bergson, H. (2007 (1923)). *The creative mind: An introduction to metaphysics* 1923. Dover Publications: Dover.

Bergson, H. (1911 [1907]). *Creative evolution*. New York: Henry Holt and Company.

Bergson, H. (1908). *Bergson to William James*. May 9, 1908.

Bergson, H. (1920). Brain and thought: A philosophical illusion from mind energy. In: *Lectures and essays* (trans: Carr, H.W., pp. 231–255). New York: Henry Holt and Company.

Bergson, H. (2001 [1889]). *Time and free will: An essay on the immediate data of consciousness*. New York: Dover Publications.

Bergson, H. (1977[1932]). Two sources of morality and religion (Trans. R. Ashley Audra & Cloudesley Brereton). Notre Dame: University of Notre Dame Press.

Bor, J. (1990). *Bergson en de onmiddellijke ervaring*. Meppel: Boom.

Brillouin, L. ([1956, 1962] 2004). *Science and information theory*. Mineola, NY: Dover.

Broca, P. (1861). Remarques sur le siège de la faculté du langage articulé, suivies d une observation d aphémie. *Bulletins et Mémoires de la Société Anatomique de Paris, 2*, 330–357.

Canguilhem, G. (1947). Note sur la situation faite en France à la philosophie biologique. *Revue de Métaphysique et de Morale, 52*(3), 322–332.

Colbrook, C. (2014). *Death of the PostHuman: Essays on extinction* (Vol. 1). London: Open Humanities Press.

Damasio, T. (1994). *Descartes error: Emotion, reason and the human brain*. New York: Avon Press.

Deleuze, G. (1991 [1966]). *Bergsonism* (trans. Tomlinson, H., Habberjam, B.). New York: Zone Books.

Einstein, A. (1934). *The world as I see it*. New York: Covici Friede.

Engel, F. (2004). Psychology and metaphysics from Maine de Biran to Bergson. In *The history of mind*. Springer.

Flourens, M. J. P. (1824). *Recherches expérimentales sur les propriétés et les fonctions du système nerveux, dans les animaux vertébrés (ed 1)* (Vol. 26, p. 20). Paris: Chez Crevot.

Fuster, J. M. (1995). *Memory in the cerebral cortex: An empirical approach to neural networks in the human and nonhuman primates*. Cambridge: The MIT Press.

Jacobson, M. (1993). *Foundations of neuroscience*. New York: Plenum Press.

Malabou, C. (2008). *What should we do with our brain?* (transl. from French by Rand, S.). New York: Fordham University Press.

Massey, H. (2015). *The origin of time: Heidegger and Bergson*. Albany: Suny Press.

Middleton, D., & Brown, S. D. (2005). *The social psychology of experience. Studies in remembering and forgetting*. Thousand Oaks: Sage.

Moeller, K., Willmes, K., & Klein, E. (2015). A review on functional and structural brain connectivity in numerical cognition. *Frontiers in Human Neuroscience, 9*, 227. https://doi.org/10.3389/fnhum.2015.00227.

O'Regan, K. J. (2011). *Why red Doesn t sound like a bell: Explaining the feel of consciousness*. Oxford: Oxford University Press.

Petit, J. L. (1997). *Les neurosciences et la philosophie de l'action*. Paris: Vrin.

Ribot, T. (2005 [1881]). Les maladies de la mémoire, Éd. Germer Baillière, coll. «Bibliothèque de philosophie contemporaine» (1881), dernière réédition 2005.

Ribot, T. (2001 [1885]). Les maladies de la personnalité, Félix Alcan, coll. «Bibliothèque de philosophie contemporaine» (1885), dernière réédition 2001.

Schopenhauer, A. (1960 [1839]). *On the freedom of the will*. Oxford: Basil Blackwell.

Thompson, E. (2007). *Mind in life: Biology, phenomenology, and the sciences of mind*. Cambridge, MA: Harvard University Press.

Tymieniecka, T. (Ed.). (2009). *Memory in the Ontopoiesis of life, book 1, memory in the generation and unfolding of life* (Analecta Husserliana, vol. CI). Dordrecht: Springer Verlag GmbH.

Wernicke, C. (1874). *Der Aphasische Sysmtomenkomplex*. Breslau: Cohn and Weingart.

Wong, A. (2014). *Life and technology in Bergson's creative evolution*. http://hdl.handle.net/2268/186268. Accessed 13 Oct 2019.

Yildirim, F. B., & Sarikcioglu, L. (2007). Marie Jean Pierre Flourens (1794–1867): An extraordinary scientist of his time. *Journal of Neurology, Neurosurgery, and Psychiatry, 78*(8), 852.

Chapter 6
Interlude: From Epistemic Debate to an Ethics of Technology

In these past chapters I have given an account of the relevance of the philosophical oeuvre of Henri Bergson by first contextualising Bergson's views with regard to three scientific domains – physics, life sciences, neuroscience – and subsequently by elaborating ton Bergson's ideas. Finally, I pointed to some elements that still seem relevant for these domains in the current situation. This, however, does not mean that these domains must be seen as separate, compartmentalised blocks. Interrelated discussions, perspectives and concepts apply to all three domains, and form a series of interwoven threads. The question that precedes all other questions is not the question of the self, nor that of matter, nor that of life. It is the question of how to account for 'one' and 'manifold': how are divergence, individuation and discreteness possible whilst direct, immediate experience is always in flux?

Approaches to the philosophy of science differ in their view on the emergence of knowledge systems. Some consider knowledge to be symbolic and abstract in nature, others rather see knowledge as something organic. It is significant to know that Bergson's father, Michal Bergson (1820–1898), was a composer who promoted the music of Frederic Chopin. The meandering melodies of Chopin are typically romantic in style, and influenced a number of later composers, Claude Debussy amongst others, whose compositions were quite familiar to Bergson. In a 1910 interview, he admitted having an 'intuitive predilection' for Debussy's music (Aimel 1910). If one were to compare philosophers to composers, one might say that Gottfried Wilhelm Leibniz's philosophical ideas bear resemblance to the compositions of Johann Sebastian Bach, while Henri Bergson's philosophy bears resemblance to those of Claude Debussy. Leibniz' monadology is, like the compositions of Bach, an intricate building consisting of complex, interlocking topological structures, whilst Bergson's process thinking is like the compositions of Debussy in its treatment of processes of convergence and divergence in time.

The tension between continuity and discreteness is one of the deeper lying and more abstract subjects in the works of Bergson. Whilst discreteness can be experienced immediately, and whilst the intellect is able to grasp the nature of discreteness, it does so in terms of a separation into distinct 'places' in space. Continuity, in

© Springer Nature Switzerland AG 2021
L. Landeweerd, *Time, Life & Memory*, Library of Ethics and Applied Philosophy 38, https://doi.org/10.1007/978-3-030-56853-5_6

terms of temporal flux, cannot be grasped in terms of space. It remains problematic from a worldview that reduces duration to geometry and mechanism. It is, as a result, reduced or excluded from many theoretical and scientific positions. This may be a result of the natural predisposition of the intellect, which is spatially oriented rather than process-oriented. For a process philosopher like Whitehead, there is no continuity of becoming, but rather a becoming of continuity. Here, Bergson retains a strong position against intellect, prioritising intuition, whilst Whitehead retains a prioritisation of the intellect.

We may be sensory conditioned to a Cartesian worldview. Due to the structure of our language, we tend to think in terms of a separation of self (first-person perspective) and world (third-person perspective). The dominance of the sense of vision also emphasises this tendency, and is emphasised by this tendency in turn. As I remarked in the first chapter of this book, musical thinking, related to sound and melody, is structured quite differently. The fact that Bergson's father was a composer may well have contributed to a higher sensitivity to this different way of thinking. The dominance of vision (seeing the world from a first-person perspective) also implies that we tend to perceive the world in terms of a collection of things. This is what the Cartesian 'mistake' amounted to: thinking about the world in terms of an analysis of spatially extended objects, parts of those objects etc. Ultimately, in understanding the world in terms of substances, rather than in terms of processes, change, continuity and fluidity. The tension between continuity and discreteness is, in this regard, one of the deeper lying and more abstract themes in Bergson's work.

Our inability to 'think change' is based in the way our intellectual ways of knowing the world tend to follow a dualistic path: 'here' are 'we', and 'there' is 'the world'. Without going too deeply into his linguistic ponderings, Martin Heidegger was indeed right to diagnose that the restricted nature of established knowledge forms was determined by language. Whilst Bergson mostly shied away from the thorny issue of the symbolical, the linguistic, and the conceptual – as discussed in relation to Čapek's interpretation of Bergson, specifically with regard to the problem of the universals – the section he did devote to this topic carry along a similar conclusion: the nature of our language predetermines, at least in part, the conclusions that we draw concerning the nature of reality.

There is an important objection against the view of the ordered universe, however: there is no such thing as a 'clean' process. The idea of a universe consisting of clean processes is an approximation that only highlights one particular tendency of the universe. Everything regarded as 'disturbance' actually forms an extremely important part of reality: the part from which chaos seems to erupt, but giving rise to complexity as well, and, on this basis, to creative evolution. The more complex the system is, the more this tendency reveals itself – but it is present in any system. Processes materialise time and display properties of self-organisation. Now, the tendency of those adhering to a deterministic view underlying classical mechanics will emphasise that all these processes, at the basis, still follow the universal laws of physics (deterministic, mechanical etc.). But those adhering to complex systems thinking or chaos theory may respond with, again, an account of the slight disturbances within those processes, deeming them more fundamental, etc..

In this post-Kuhnian age, one might do away with these incompatible accounts of the universe as merely two opposing discourses about a reality that essentially remains beyond our linguistically conditioned event horizon. But from a (pan)vitalist position, language itself might be considered as a life form – something Wittgenstein pointed out in his later works. And as such, it would not be isolated from the complex life forms surrounding it, nor is it merely a parasitic (or epiphytic) form of life. It is here that the concept of the virtual and the imaginary enter into the debate. It is also here that the distinction between 'going around' an object – studying it analytically from different angles –, and 'entering into' it becomes relevant.

Mechanist theories in the exact sciences proceed on the assumption that the world consists of substances that are ultimately clearly differentiated: the basic premise is one of a world consisting of pure and clearly circumscribed material things. The number of a species, although big, is finite. Water, as H_2O, consists of one particular type of molecule. Nerves either signal a response or not. But reality is never pure. It is always made up of mixtures and nuances. Although the artificial realities construed by mechanist theories are ontologically deficient, they do yield valuable practical knowledge. But it is a type of knowledge that is always, by necessity, approximate. Aqua destillata is never pure H_2O, the exact number of individuals belonging to a species is impossible to determine with certainty, and although I currently do not experience any physical discomfort, the ache in my elbow is caused by a cluster of nerves, some of them dormant, some active and some somewhere in between. This also means that although bridges, aeroplanes and food preservatives may have become a lot safer due to the increase in scientific knowledge concerning materials, aerodynamics and contamination, the worldview behind these developments leads to an overconfidence in the certainty of its outcomes. And this whilst the order of the 'real' is typified by unpredictability.

Like any theoretical framework, mechanist theories in the exact sciences tend to have a filtered idea of empirical reality. They proceed on the assumption of clearly definable substances. They depart from the ideal of purity, and arranges its method around a conditioning of the behaviour of the phenomena studied in isolated and purified states. As Gaston Bachelard stated, the scientific endeavour consists of not studying nature as it presents itself to us on its own accord, but a highly artificial distillation of nature (Zwart 2019). The laboratory functions as an environment of purifies natural phenomena. This is the basis of the empirical method. It has become one of the two keystones of modern science. The other keystone is quantification. The assumption is that to be able to grasp the fundamental behaviour of the universe and its phenomena one needs to make this behaviour calculable. This implies that all phenomena have, in one sense or another, a definitive nature.

The mechanistic worldview dismisses the white noise as disturbance, to focus on the classifications, definitions and causal relations it attempts to establish. This is problematic since this white noise is not just the rubble, the disturbance, or the sand in the cogwheels of the machine; it is a driving force in reality itself.

There are two very distinct forms of time at play in this incommensurability: the geometrical time of the mechanical thinkers (from Parmenides and Zeno via Newton up to Einstein) and the living time of the organic thinkers (the time of Tao and Heraclitus, but also of Bohm and Bergson). The geometrical conception of time is actually a non-time, since it takes the concepts of space and geometry and applies them to time. If Bergson is correct in stating that time is of a fully different order, conceiving of time as a spatial category (a fourth dimension, or a succession of moments, represented by an infinite number of points on a line) leads to misrepresentation. It leads to a characterisation of change on the basis of concepts that can only describe fixed and unchangeable relations. Complexity cannot be reconciled with mechanist thought, since mechanistic thought obfuscates time, and therefore the notion of complexity. Time cannot be accounted for without an irreducible notion of change.

Bergson revisited a painful issue related to the neo-Kantian reduction of Kant's epistemological architecture of our knowledge system to the structure of the brain. Neo-Kantian philosophers believed that our ways of knowing were wired into the structure of our brains, and were therefore materially determined (see Chap. 5). Although refraining from material reductionism, Bergson did agree with the view that we are psychologically predisposed to perceive the world in terms of causality. In 1900, at the first International Congress of Philosophy, he presented an important paper entitled 'On the Psychological Origins of the Belief in the Law of Causality' (Sur les origines psychologiques de notre croyance à la loi de causalité: Bergson) (Bergson 1900). In this paper he defends the idea that although we are predisposed to think the world in terms of cause and effect, this psychological predisposition is not inescapable. It is lodged in the structure of our ways of thinking as well as the structure of language, but it does not exclude the possibility of alternative ways of thinking. It does however put a strain on our use of language. Since the concepts that we are accustomed to are embedded in the structure of our language, finding alternative ways of thinking by necessity involves an intuitive grasp rather than rational analysis. Time is in need of other words, another vocabulary. It is for this reason that Bergson had, as Čapek also discusses (1971), such a distrust in language. Our tendency towards substance-ontology, thinking the world as a collection of things, is founded in the subject-predicate-structure of language. Only the language of poetry should be trusted (van Dongen 2014, pp. xx), since it brings words in rhythm and melody rather than structuring thought analytically.

In *Creative Evolution* (1907), Bergson considers life in terms of temporal movement, defined as duration as we have seen. This view on time is presents an alternative to the mechanistic view of time in terms of predetermined cause and effect. The temporal movement that life is, is retained in memory. At the same time, life also has an extensive nature: it takes place in space, and here the practical aspects of our organic existence can be identified. Thus, we endure through time, and relate to this enduring existence through memory, whilst we also have an identity, in which we are singular and not continuous with the rest of creation. This extended and discontinuous nature of what we are, is shaped in terms of the practical qualities of our bodies. The two are interrelated but the one aspect cannot be reduced to the other, nor vice versa. Their interrelation can be examined by examining change. Vital impulse, the defining property of living structures, is common to all of life, and whilst one might state that even inorganic matter shows some indications of such an impulse, it is only really apparent in the complex variability of converging and diverging living structures. This variability over time demonstrates a tendency that is captured by Erwin Schrödinger's later definition of life as negative entropy or, to use Leon Brillouin's term (1956) as 'negentropic': the self-organisational tendency of life that seemingly counteracts the principle of entropy in thermodynamics. Bergson defines this negentropic tendency of evolution as creative.[1] It also manifest itself in the emergence of instinct and, eventually, intelligence. Instincts are already observable in lower species such as invertebrates, but Bergson also uses the term instinct with regard to plant life. Instinct informs both simple and complex behaviour. Intelligent behaviour, in his view, can only be observed in vertebrates and notably in mammals. Intelligence is the ability to approach the extended world in a creative manner. But our intelligence also incites us to analyse the temporal, enduring aspects of our world in spatial terms. As humankind is the most intelligent creature in the world, we run into a paradoxical situation. Although intelligence is superior to instinct, it has the tendency to obfuscate duration. Therefore, intuition is important as a third human ability, allowing us to intuit the more dynamical aspects of the world and place ourselves within the creative flow of vital impulse. Only through such an intuitive approach can we overcome the difficulties and the dualistic tendencies of rationalistic approaches to metaphysics.

In view of its emphasis on duration and on vital impulse, Bergson's philosophy can be subsumed under the tradition of process metaphysics. Process metaphysics, or process thinking, is an ontological approach which understands being in terms of change. The more dominant approaches to ontology in western philosophy either deny change, or see it as a secondary phenomenon in a reality that is basically composed of unchanging substances. Thus, physical movement is analysed in terms of the combination of three-dimensional space and linear time, while life is seen as composed of non-living components and spirit is analysed in terms of fixed neurological pathways and their biophysical properties. This seems to suggest that, ultimately, knowledge and change are incompatible. In conceptual knowledge we attempt to think matter. But in this framework, we cannot really comprehend matter due to its focus on how things on the macro-scale are composed of smaller components. George Canguilhem disagrees with Bergson in this respect, as he believes that duration can be thought conceptually, whilst Bergson perceives a barrier between intelligent rational thought and the ever-changing nature of life, due to the

---

[1] As such 'creative negentropy' (Gunter 2013) might be seen as the best circumscription of the more elusive 'élan vital' (or vital impulse).

intellectual nature of conceptual language. As such, process thinking itself is ultimately insufficient to reconcile world and thought, life and knowledge. Intuition on the other hand, as an immediate form of knowledge, is brought about by duration itself, and can thus revitalise our knowledge of life.

Where one would expect a thinker such as Bergson to be suspicious of any aspect of technology, the opposite is true. Technology, in Bergson's oeuvre, is not contrasted with biology, but is rather seen as a function or extension of biology and, by consequence, as an inherent part of evolution. At the same time, technology, as a tool of human intelligence, distances us from a direct experience of the world. This tension between immediate experience – data of consciousness – on the one hand and Bergson's ambiguous understanding of technology on the other, is addressed in *The Two Sources of Morality and Religion* (1932).[2]

A superficial reading of Bergson's ideas might suggest a binary categorisation that divides reality into different realms to which different rules apply: the movement of inorganic objects determined by natural law and the behaviour of plant and animal life forms spurred by impulse and instinct (growth, evolution, expansion). Finally, we arrive at human existence as supposedly free, due to humankind's faculties of memory and imagination. Such a compartmentalisation of the world is problematic for various reasons. It divides reality into different ontologies and that is not what Bergson meant to do. In a sense, all these different areas demonstrate durations and vital impulse to some extent. The only distinction is the level of aggregation on which these tendencies are at work (movement, instinct, memory and imagination). Still, there is something special about human experience and this has to do with imagination. Bergson adheres to a non-materialist view on memory. For Bergson, memories should be understood in terms of their relation to time, and not reduced to the grey mass inside our skulls.

Literature

Aimel, G. (1910, décembre 11). Une heure chez Henri Bergson, par Georges Aimel, Généralités sur le Cours du Collège de France consacré à la personnalité. In: *Mélanges (1972)* (pp. 843–844). Paris: Presses Universitaires de France.

Bergson, H. (1900). Notes sur les origines psychologiques de notre croyance à la loi de causalité Sur les origines psychologiques de notre croyance à la loi de causalité. *Bibliothèque du Congrès International de Philosophie, 1*, 1–15.

Bergson, H. (1911 [1907]). *Creative evolution*. New York: Henry Holt and Company.

Bergson, H. (1977 [1932]). *Two sources of morality and religion* (trans. Audra, R. A., Brereton, C.). Notre Dame: University of Notre Dame Press.

Brillouin, L. ([1956, 1962] 2004). *Science and information theory*. Mineola, NY: Dover.

---

[2] His last work was a collection of essays that included his 1903 introduction to metaphysics, published in French in 1934 under the title *La Pensée et mouvant*, and in English under the title 'The Creative Mind'. All nine articles included had however already been published previously, between 1903 and 1923.

Gunter, P. (2013). Bergson and Proust: A question of influence. In P. Ardoin, S. E. Gontarski, & L. Mattison (Eds.), *Understanding Bergson, Understanding Modernism*. New York: Bloomsbury.

van Dongen, H. (2014). *Bergson*. Amsterdam: Boom.

Zwart, H. (2019). Iconoclasm and imagination: Gaston Bachelard's philosophy of technoscience. *Human Studies*.

Chapter 7
To Become Gods, or to Perish in the Process…

> *Mankind lies groaning, half-crushed beneath the weight of its own progress. Men do not sufficiently realize that their future is in their own hands. Theirs is the task of determining first of all whether they want to go on living or not. Theirs the responsibility, then, for deciding if they want merely to live, or intend to make just the extra effort required for fulfilling, even on their refractory planet, the essential function of the universe, which is a machine for the making of gods*
>
> *–Henri Bergson, The Two Sources of Morality and Religion*

When Bergson was awarded the 1927 Nobel Prize for literature, his health was too fragile to travel to Stockholm for the award ceremony. Eventually, in 1928, Armand Bernard, a French diplomat ('ministre plénipotentiaire') living in Stockholm, accepted the award on his behalf and read the speech that had been drafted by Bergson. In his speech, Bergson discussed the scientific and technological optimism ingrained in the concept of the Nobel Prize. He voiced a critical position. The assumption that the industrial revolution had brought social progress was to his mind naive: "If the nineteenth century made tremendous progress in mechanical inventions, it too often assumed that these inventions, by the sheer accumulation of their material effects, would raise the moral level of mankind" (Bergson 1967 [1927a]). The view that technological invention and progress would result in a progress of civilisation, morality and the intellect was the assumption of the nineteenth century positivists: discovery and invention as key to social progress and progress of civilisation. And although strongly criticized both within and outside philosophy, it is still a position many adhered to today. The very word 'innovation' carries along connotations of not merely new but automatically also better. It is the assumption of infinite progress: we come up with ever smarter ever more practical and ever more effective technologies. They will first of all solve issues connected to world poverty and hunger, pollution, disease and climate change. Then they will be used to serve humankind and create the world we desire, rather than merely inhabiting the one we got thrown into.

© Springer Nature Switzerland AG 2021
L. Landeweerd, *Time, Life & Memory*, Library of Ethics and Applied Philosophy 38, https://doi.org/10.1007/978-3-030-56853-5_7

In 1932, Bergson published his last major work *The Two Sources of Morality and Religion*. It outlines a basis for a morality that is not restricted to specific regions, cultures or communities. This next chapter gives an account of the relevance of Bergson's thought for a diagnosis of the present. Rather than seeking to give an account of the history of technology, then the relevance of Bergson's works for its analysis, to the extrapolate to the present, I have structured this chapter with a specific focus on the nature of the relation between humanity and technology. Bergson's *The Two Sources of Morality and Religion* (1932) forms an introduction to what would become modern philosophy of technology. Seen the relevance of many authors from this field and their direct or indirect relation to Bergson, this chapter will discuss the relation between humanity and technology through the eyes of a broad tradition in philosophy, rather than merely those of Bergson.

Invention is often treated as good in itself. In the policy language of the European Union, innovation has become a term that replaces the paired terms 'science and technology' as well as the earlier terminology of 'discovery and invention'. This shift in discourse signifies a shift in political orientation. Replacements of names and terminology are hardly ever innocent (Zwart et al. 2014). And in this case, it reflects a more basic shift of the European Union from a politics of internationalism to a politics of economic growth. It is strongly interconnected with the shift towards a neo-liberal political agenda, in many member states of the European Union, but also in the European Commission and Parliament.

Technology, within this new and optimistic perspective, is presented as a cornucopia; a horn of plenty that will ultimately be able to provide us with our heart's desires.[1] The view on technology as a cornucopia for society also has a darker side however: Pandora's box. Two containers, of which the one is a symbol for the bountiful fruits of nature, the other a symbol for natural disasters. Whilst the cornucopia was a relic of Zeus's divine strength, who wrestled the horn from a giant goat's head, Pandora's box was given by Zeus as a deceitful present to Epimetheus and Pandora, his wife to be. When she opens it, in spite of being warned, all the evils and miseries of the world fly out to plague mankind.[2] In this chapter I will discuss the relevance of Bergson ideas for the philosophy of technology.

In the text for his acceptance speech, Bergson outlines a view on machines as artificial organs that extend the scope and functionality of our natural organs. This view would be elaborated in his later publication *The Two Sources of Morality and Religion* (1932). In this view, we should consider technology as an enlargement of the body of humanity as such. Bergson's concern is that, although our bodies' artificial extensions have grown exponentially, our souls have remained in their original state, so that the gap between our expanding technological and our restricted moral

---

[1] This horn was taken from the head of Amalthea, the goat that nourished Zeus when he was hidden in his infancy, and as the goat provided nourishment for the God, the horn provided produce (in the shape of nuts, fruits, vegetables and flowers) for mankind. This symbol that originally applied to agriculture gradually came to refer to any form of abundance or prosperity.

[2] In *Technics and Time 1: the Fault of Epimetheus* (1998), Bernard Stiegler takes this myth as central for understanding technology. In this chapter I will reiterate this myth.

power increases. This is the cause of a number of social and political problems. Thus, the emergence of trains, cars and telephones was expected to decrease not only the spatial distances between peoples, but also the separations between their moral systems, cultures and beliefs. But Bergson states that rather the opposite has been brought about by these inventions.

In Bergson's view, the ideal of the Nobel Prize Foundation was based on the idea that the human intellect might one day opt for a single and homogeneous 'republic of minds'. This optimistic outlook on the rational intellect as a basis for moral unification stems from an earlier age, the age of Enlightenment. But it needs more than intellectual effort to safeguard a 'moral rapprochement' between the diverse life forms of human civilisations.

Since Bergson's acceptance speech, almost a century has elapsed. The invention of machines has, from the 1920s until now, expanded dramatically, and it has taken a shape that could not have been anticipated at the beginning of the twentieth century. The invention of the internet has made communication and sharing of information possible to an enormous degree. Here, it appears that Bergson's moral criterion for invention – to enable a 'rapprochement' between peoples and cultures – was met. But the communicative potential of the internet also seems to nurture new antagonisms between groups and peoples. Social media, presented at one stage as an enabler of the Arabian spring movements, also enabled the emergence of radical and violent voices and groups.

To fully understand Bergson's attitude towards technology and his influence on others, we should make a distinction between Bergson's criticism of mechanistic thinking on the one hand, and Bergson's views on technology on the other. Mechanistic thinking is incapable to understand the true nature of life. This is, as we have seen, the core motif of *Creative Evolution* (Bergson 1907 [1896]) because of its tendency to compartmentalise human experience into separate parts. This blocks a valid understanding of concrete time, in terms of stream, flux and continuity. It cannot provide an understanding of the flow of duration in the abiotic physical world, the process of evolution in life, and the stream of thoughts in our flow of consciousness. The intellect, with the faculty of rational understanding, cannot grasp these aspects without reducing them to discrete elements outside of our intuitive horizon. Technology, however, is not a way of understanding the world, but rather a product of life. (Image 7.1)

Technology is the manifestation of intelligence in evolution, and as such deeply intertwined with the emergence of the human being (Bergson 1907 [1896]). In *The Two Sources of Morality and Religion* (1932) Bergson even attributes a specific kind of mysticism to what he phrases as 'mechanistic invention'. He considers technology as a supplement of our souls, since only mechanical invention can mediate between the pull of matter and the force of spirit and morality.[3] Matter, for Bergson,

---

[3] Here, again in spite of himself, Bergson echoes an observation already made by Immanuel Kant: in his oeuvre, Kant described three different uses of reason: technical, pragmatic, and moral. They inform respectively skill, prudence, and morality. In his 'Conjectural beginnings of human history' (1786) Kant describes their manifestation as the beginning of civilisation. Speculating that the development stage of agricultural technologies necessitated a different role responsibility to earlier stages of development such as nomadic shepherds, moral reasoning presupposes pragmatic reasoning (a notion of prudence), which in its turn presupposes technical reasoning (skill, ability). So also for Kant, technology precedes and is preconditional to morality.

Image 7.1 Living root bridge, Meghalaya, India

becomes a precondition for morality, which would not exist if our relation to the world was mere direct experience.

Bergson's views on technology were taken up by Pierre Teilhard de Chardin (1881–1955), preluding a generation of French philosophers that focus on the analysis of technology as a complex social system. This includes philosophers such as Gilbert Simondon, George Canguilhem, Michel Serres, Jacques Derrida, Bruno Latour and Bernard Stiegler. An important author in this tradition outside France is the Indian philosopher Sri Aurobindu (1872–1950), an Indian guru and philosopher who joined the Indian independence movement. His writings concern specific views on the nature of human and posthuman evolution, and the relation between humans and technology.

Pierre Teilhard de Chardin was a Jesuit priest, philosopher and renowned palaeontologist whose ideas were deeply influenced by the philosophy of Bergson. Whilst his works were received with enthusiasm by Catholics dissatisfied with the contrast between church and science, his works have been somewhat neglected over the past decades. His process metaphysical account of technology include a view of technology as part of a broader, cosmic evolutionary tendency, a resistance against teleological and substantivist accounts of the world and a criticism of mechanistic thought. It also includes a philosophical vitalist view on how life emerged from abiotic matter, and how consciousness emerged out of living nature. This carries with it a specific view on the nature of technological systems and the nature of technology innovation.

The uniqueness of human existence is also elaborated by Bergson in the political and cultural realm, notably via his work for the Commission internationale de coopération intellectuelle (International Committee on Intellectual Cooperation, or CICI, the pre-war predecessor of UNESCO). For indeed, although so far, we have discussed Bergson's views regarding the domains of physics, the life sciences and neuroscience, his convictions eventually cumulate, one could argue, in his views on and practical role in international politics. In 1921, Bergson was invited to become the first chairman the CICI. The aims of this organisation was to nurture international exchange between different traditions, cultures and communities. On the basis of his work for this organisation,

Due to the fact that technology extends our bodies beyond their biologically evolved functions, it has become too large for our souls to grasp. In this regard, one might consider technologies as a 'physical swelling'. Progress, from this perspective, is not a move forward, but an expansion outwards, as Simonon-scholar Pascal Chabot sketches it (2013 p. 148), now spanning the globe. It is in his work for the League of Nations (the predecessor of the United Nations), more specifically for the CICI that Bergson first saw the need for a new, more global and inclusive approach to morality that would fit the expansion of our bodies through technology. In short, Bergson saw a need to move beyond the evolutionarily informed static morality of the group, the clan, the tribe.

Bergson ties technology in with morality and mysticism. Technology is a mystical moment because it is through technology that creative evolution becomes a truly global process: a theme addressed by Teilhard the Chardin as the emergence of the noosphere (Zwart 2016). This thematisation of technology and mysticism should not come as a surprise. Rather, the problem is that we are no longer aware of this inherent mystical dimension of technology. Vladimir Jankélévitch, a younger pupil of Bergson, argued that "if our bodies are enhanced and swell up through technology, we cannot expect them to be guided by a virtuous soul". There is no virtue to be expected from this 'extended musculature', as Jankélévitch called it (2008 [1931]). In this regard, it would be naive to hope for a more elevated morality. Instead, we should turn to more simple ways of living and experiencing, through music, through joy, but not through technological culture.

Gilbert Simondon (1924–1989) was a philosopher of technology inspired by Bergson. Simondon discussed the Bergsonian views of becoming, emergence and creative evolution in the context of cybernetics, chaos theory and complex systems theory. His philosophy encompasses an understanding of processes of individuation that replaced the ontology of things by an ontology of processes. His views also imply that one can never truly discern an individual thing because everything is in a continuous process of becoming (rather than *being* something). Where Jankélévitch was critical of the hope expressed by Bergson on technology, Simondon was critical of Bergson's allusion to spirit, mysticism and intuition as privileged over intelligence, technology and matter. Bergson conceives of the relation between intelligence, matter and technology as a relationship in which intelligence imposes itself on matter by creating technological artefacts as, in Chabot's words, "a game played with modelling clay or Lego bricks" (Cabot 2013, p. 149). As such, matter is both a vehicle and an obstacle for the materialisation of intelligence (Bergson 1932, p. 94). Here the parallel to the metaphors used in fields of bioengineering such as germline modification and synthetic biology can hardly go unnoticed, for indeed, Lego bricks are used as the explanatory device par excellence in this field. I will return to these metaphors in the second paragraph of this chapter, but first I will turn to a problem area induced by our extensive use of technology, for purposes of transportation, heating, cooking, etc. on a scale that spans the globe.

Bergson's ideas seem especially relevant in view of the current convergence of different scientific disciplines with each other and with technology, innovation, and the impact of technology on societies globally. This involves on the one hand a growing human responsibility for the planet, in terms of climate change, loss of biodiversity and pollution; on the other hand, the very nature of being human is at stake via our increasing ability to self-adjust and self-create. We are evolving (as a recent step in the process of evolution) from a being which merely inhabits the surface of this planet together with may other species into a being that takes responsibility over nature – including its own nature. This necessitates a collective mentality shift, but moral progress is a much slower and difficult process than technological progress.

To understand the importance of this theme for Bergson, we should pay attention to a persistent theme in his oeuvre: his criticism of mechanistic thinking. Over the years between the publication of *Matter an Memory* (1896) and *The Two Sources of Morality and Religion* (1932), Bergson took different positions towards mechanicism (see figure below).

Publication	Basic position to mechanics	Explanation
Matter and Memory	Mechanicism versus direct experience	Mechanicism as the basis for locationist perspectives on human mind
Creative Evolution	The mechanic versus the organic	Mechanicism as a static interpretative framework for issues of life, resulting in an inability to distinguish between life and non-life
Duration and Simultaneity	Mechanicism as responsible for the dimensionalisation of time	Mechanical thought as a reductionism of concrete time to spatial categories
The Two Sources of Morality and Religion	Mechanisms as mystical in nature	Mechanicism not as a faculty of the intellect but as an expression of the human ability to transcend the confines of his naturally evolved organs[4]

Bergson first framed mechanicism as the main culprit in the misunderstandings of time, life, and memory. But he did not treat technology as synonymous with mechanics. In *Creative Evolution* he considered mechanics as the natural complement of the organism, and machines as the extended organs of life. So, already in 1907, Bergson emphasised the more positive potential of technology. Mechanics invokes mysticism, and mysticism invokes mechanics (Bergson 1932, p. 330). Why Bergson saw these two polarities as intertwined will be the core issue of this

---

[4]These views were the first steps towards an organological philosophy, proponents of which include Georges Canguilhem and, more recently, Bernard Stiegler.

chapter. I will attempt to explain this by giving an interpretation of the enigmatic quotation that opened this chapter.

Section 7.1 will discuss the problems associated with the – somewhat fashionable – term 'Anthropocene'. In Sect. 7.2, this will be connected with an alternative view on the emergence of technological systems. Ultimately, in Sect. 7.3, building onto the findings of both the preceding sections and the preceding chapters, a proposal will be elaborated for a moral philosophy of science that might be termed a 'post-natural' or 'post-evolutionary' ethics.[5] It supports a shift in the discussion on current approaches to the ethics of research and innovation, both in their institutional and their pragmatic setting.

7.1 Imagining the Anthropocene

"For its World Heritage Centre in Paris", I wrote after a visit to UNESCO Paris, "the UNESCO's committee of architecture and art commissioned a tableau from Pablo Picasso. The tableau, which was unveiled in 1958, can still be found in an irregularly shaped part of the building, covering a wall that separates a large hallway. It was initially given the title 'The Forces of Life and the Spirit Triumphing over Evil'. Later, the 40-tile tableau came to be known as 'The fall of Icarus'. Picasso's tableau depicts distorted figures in different poses near a sea, surrounding a charred figure falling from the sky. In its initial reference to the triumph of life and spirit over evil, *The Fall of Icarus* signifies a struggle between the forces of good and evil. Mankind is both languid spectator and dramatic victim in this scene: Picasso himself suggested the tableau merely depicts "des gens qui se baignent, tout simplement"; a simple scene of bathers. What once was intended as a dark spirit, a fallen angel, became an overconfident human who flew too close to the sun. But his dramatic fall into the sea is witnessed with awkward indifference. Our generation is confronted with climate change, loss of biodiversity and large-scale pollution of our ecosystems. The scars will be of such a scale that they will remain permanently visible in the earth layers currently formed as well as in dramatic shifts in animal and plant life. These shifts signify the birth of a new geological era: the Anthropocene. Sustainable development is often suggested as a key strategy to counteract the negative impacts. But as Picasso's bathers, we hardly bother to look up from our daily affairs to this disaster in slow motion." (Landeweerd 2018). (Image 7.2)

In reference to the charred central figure of Picasso's Icarus, Hub Zwart (2016, p. 77) states: "Picasso's X-ray figure exposes humanity's genetic 'essence' (DNA). Nuclear bombs and the discovery of DNA, as landmark 'achievements' of twentieth-century science, brought about by elementary particle physics and molecular

---

[5] Here, it should be taken into account that strong arguments exist in theoretical debates over the nature of ethics, that an evolutionary ethics is, strictly speaking, impossible. The discussions in question relate to both David Hume's *A Treatise on Human Nature*, and G.E. Moore's *Principia Ethica* (1903).

Image 7.2   Picasso, The Fall of Icarus, 1958. Acrylic on forty wooden panels. 910 × 1060 cm

genetics, respectively, converged in disclosing the letters (Στοιχεῖα) of matter, energy and life, thereby obliterating the living, which explains why physicists (Delbrück, Schrödinger, Wilkins, Crick, etc.) played such a pivotal role in the post-war molecular biology revolution." Here, Zwart points to an essential issue: the translation of life into code, letters of an alphabet (C, T, A, G) as it was inspired by, amongst others, Erwin Schrödinger. But in this translation, all sight is lost on systemic aspects such as change, evolution, emergence, etc. whilst these define life. Without these aspects, what is central to life – that it is living – is obliterated.

Where the living complexity that surrounds us has evolved over large timescales, the technological systems that emerged in the course of our own evolution are now developing at an accelerating pace. This means that evolution is no longer restricted to the natural, DNA-based forms of variation and the selective processes of environments that change life forms at a slower pace. But in our collective behavioural patterns on the other hand, this step beyond traditional evolution has been accelerating mainly on the basis of the uncovering of energy resources that were accumulated during extremely slow processes; oil, natural gas, coal.

Mankind's impact, resulting in dramatic global change at an accelerating pace, amounts to an extreme geological crisis, comparable to the impact of a large meteor on the planet's surface. We hardly realise the fragility of our planet's living surface in the vastness of space and the absolute exceptionality of our existence on its surface. As Schopenhauer wrote at the beginning of book II of *The World as Will and Representation*: "In endless space, countless luminous spheres, round each of which some dozen smaller illuminated ones revolve, hot at the core and covered over with a hard, cold crust; on this crust a mouldy film has produced living and knowing beings: this is empirical truth, the real, the world. Yet for a being who thinks, it is a precarious position to stand on one of those numberless spheres freely floating in boundless space, without knowing whence or whither, and to be only one of innumerable similar beings that throng, press, and toil, restlessly and rapidly arising and passing away in beginningless and endless time." (Schopenhauer 1818, p. 3).

In suit with this famous Schopenhauerian paragraph, Nietzsche, in his *On Truth and Lie in the Nonmoral Sense* wrote a more cynical sketch of our limited existence: "Once upon a time, in some out of the way corner of that universe which is dispersed into numberless twinkling solar systems, there was a star upon which clever beasts invented knowing. That was the most arrogant and mendacious minute of world history, but nevertheless only a minute. After nature had drawn a few breaths, the star cooled and solidified, and the clever beasts had to die. The time had come too, for although they boasted how much they had understood, in the end they discovered to their great annoyance that they had understood everything falsely. They died, and in dying cursed truth. Such was the nature of these desperate beasts who invented knowing."(Nietzsche 1999 [1873], p. 79).

Map makers and globe makers already imagined the earth from an outer perspective, whilst 'sophonauts' like Schopenhauer and Nietzsche from a philosophical perspective, which comes close perhaps to a God's eye view. Their imagined perspective was however not matched by direct human experience until Yuri Gagarin was sent to outer space. Gagarin and those that would follow would tell of the dramatic impact that an external view on the world had on them. This impact, also referred to as the 'overview effect', is the almost religious experience which accompanies the awareness of the fragility of the earth's atmosphere in the light of the vast, life-less space of the universe surrounding us.

Nietzsche would already refer to human existence as faulty by nature: we are, in his view 'das noch nicht festgestellte Tier' – an animal that has not yet been fixated by its nature. This notion of the human being reveals an interesting perspective on evolution: evolution as fixation. Pragmatist philosophers such as Charles Sanders Peirce and William James considered evolution as a process of fixation of species and their traits. Nietzsche's notion that we are not fixated yet carries along a view on the human being as a mere transitory moment in an evolutionary process. Our nature is something continuouslypostponed. This postponed nature, this indeterminacy is categorical, and Nietzsche sketches the indeterminacy of man as a disorder. This Nietzschean notion of human indeterminacy as our basic flaw would later inspire philosophers such as Jacques Derrida and Bernard Stiegler, using the French term 'faute d'origine'. As Nietzsche's Zarathustra, inspired by Schopenhauer's relativistic description of human existence, phrased it: "The earth has a skin and that skin has diseases; one of its diseases is called man." (Nietzsche 1885–1892 [1968]). The planet is ill, and we are the cause of the illness. Gaia's skin disease has now looked at itself, from outer space: of late, astronauts such as the Dutchman André Kuipers express their concern over the fact that the impact of humankind on the planet is becoming visible from outer space. The impact of our presence on the planet will also remain visible in the far future, reason for the Dutch Nobel Prize-winning atmospheric chemist Paul Crutzen (Crutzen and Stoermer 2000; Crutzen 2002) to announce the birth of a new geological era – the Anthropocene.

The term 'Anthropocene' has been criticised because it suggests a politicisation of the vocabulary and definitions of the geosciences (see a.o. Lewis and Maslin 2015a; Lewis and Maslin 2015b; Barry and Maslin (2016)). Geologically speaking, the definition of a breach between two epochs is based on large scale global changes that involve climate change, mass extinction and other irreversible processes. Change in geology is a slow process, hardly noticeable in media res, except for specific periods of revolution. We find ourselves in such a critical situation, and we ourselves appear to be the agent of change, rather than merely the witness of this change: this breach is caused by the technologies that we produced to allow us to transcend, in an Icarus-like fashion, natural evolution.

The postfix '-cene' derives from the Greek καινός (kainós (latinized; caenus)), meaning 'new'. Charles Lyell (1797–1875) introduced this postfix, attaching it to πλεῖον (pleion, 'more': Pleiocene; πλεῖστος, (latinized: pleīstos, "most": Pleistocene) and ὅλος (latinized: holos, 'whole' or 'entire'): Holocene. In other words, the geological eras of more new, most new and wholly new. The term Anthropocene does not embrace this etymological rule (after all, what can be the incremental step after 'more new', 'most new' and 'wholly new'). The somewhat awkward literal translation would be 'man-new'.

The Anthropocene presents us with a 'biosphere' that in effect is increasingly becoming a 'technosphere': the atmosphere is no longer a natural given, it has become a by-product of human activities. The Anthropocene confronts us with the uncanny power of science and technology. The emergence of the Anthropocene reveals the toxic character of the process of industrialisation, ruining ecological but also social, psychological, economic and cultural systems: man's influence on planetary life appears to be chaotically destructive rather than creative. Our current industrial and economic activities have given rise to a mass extinction of species and a disruptive change of the climate. The challenge ahead lies in countering these changes by adapting our socio-economic mentalities and consumption behaviours.

Ever since the age of Enlightenment a belief in infinite economic progress became a guiding conviction. This is a result of industrial and large-scale use of fossil fuels. But over the course of the twentieth century, manipulation of human behaviour became a key strategy to boost the dominant economic dogma of 'growth of growth'. Personal identity has become a key instrument: one can buy one's personal self at IKEA, define oneself through one's new car and trade one's soul in signing the user agreement for public exposure on Facebook. Still, fringe ideologies such as 'transhumanism' dream up utopias of a bright and shiny high-tech future of overabundance and enhancement of the mind and body, and the mainstream assumption is that technological progress is firmly on course. The naive assumption is that we are right on track and business as usual can continue. But we face a decline rather than an increase in welfare, whilst the limits of our planet's ecosystem-stability have been reached.

Technologies are often discussed as black boxes: you know what goes in, and what comes out, but what goes on internally – how they function – remains obscure to the non-expert, and, as a result of increasing specialisation in all branches of engineering, by now also for most experts. This box-view of technology is not only valid for the technical functions of technological artefacts, but also for the social, ecological and moral implications of technology. Here, the box can either be a cornucopia, or Pandora's bridal chest: technology can either yield anything we want or need – food, luxuries, freedom -, or anything we fear – ecological disasters, pollution, famine, subjugation. Technology is, as such, both medicine and poison, and as such, due to our apparent inability to take responsibility over its use, our favourite scapegoat.[6]

Individuals can hardly take a role in the global gambit that innovation has created regarding pollution, overpopulation, unequal distribution of wealth, knowledge and technology, protection of nature and biodiversity, exhaustion of natural resources and the limitations of planet earth. At the same time, we can no longer permit ourselves to believe that we are a comfortable part of nature. We affect the planet's health in a negative way, and this calls for a completely new form of responsibility. At a moment in the near future we will hit so called 'peak oil', the moment that we will reach the peak in our possible harvesting of fossil fuels in spite of ever-increasing technological possibilities to extract these from the crust of the earth. This peak would be reached due to the finite nature of the natural fossil fuel reserve and the still increasing global consumption rates of fossil fuels. Our current collective uses of natural resources (from extant ecosystems to fossil resources) have a grave impact on the planet's ability to sustain the current organisation of life. This calls for a new organisation of our economies, political systems and societies.

Many issues the world is confronted with can only be addressed through international coordination: waste management, climate change, loss of biodiversity, ecosystemic change, global poverty and political instability are all interconnected in kaleidoscopically complex ways. It is only through an open and dialogical approach, involving our various value systems, moral beliefs and societal differences, that we can tackle the problems that we are confronted with on a global scale. But this is not achieved easily: we cannot just depart from a universal ethics, and then 'correct' what is wrong. We are bound by material conditions, cultural habituations and social as well as political systemic conventions. Dismissing these limitations is both naive with regard to the complex interrelations between humans, technology and environment, but also disrespectful of the differences that typify the technological condition of human beings.

In Bergson's view we can only rise above 'earthly things' if we use these earthly things to wedge our efforts to self-transcend: only in relation to these earthly things, in relation to matter can we escape from matter. We need the material for the necessary leverage. Thus, in Bergson's view, the mystical is closely intertwined with the mechanical (Bergson 1932, p. 267). A problem diagnosed by Bergson was that we do not sufficiently realise this mystical nature of technology, and as a result, technology, in his time, bought comfort for the few rather than liberation for the masses. Even though times have changed, and even though such comfort is now 'enjoyed'

---

[6] Inspired by Jacques Derrida's (1981) analysis of 'writing' as pharmakon (in ancient Greek: remedy, poison, and scapegoat at the same time), Bernard Stiegler (2012) discusses technology as having such a binary nature as well.

by the masses, such enjoyment often proves to be addictive or disruptive rather than liberating.

The term Anthropocene itself calls into mind an apocalyptic sense of urgency. Geologically speaking, the definition of a breach between two epochs is based on large-scale global transitions that involve climate change, extinction etc.. Renewal in geology is a slow process, not noticeable *in media res*, except for specific periods of revolution. We find ourselves in such a breach, but now, we ourselves appear to be the agent, rather than a mere witness of this change. Most of our machine parks are still fed with fossilised fuels, accumulated over huge expanses of time and in a sense, material concretisations of lived time. Their use has enabled a conversion into (loco)motion, vast production and distribution that may have seemed a blessing at first, in terms of the resulting material successes. Technology was at first spiritual and mystic in nature because material resources (fossil fuels) we used to realise a spiritualisation of the masses, freeing humankind from material labour and opening up new worlds of intellectual labour. Yet, because we were insufficiently aware of this dimension, the de-materialising impact of technology became "distended out of all proportion" (Bergson 1932, p. 268). Our extended bodies have grown too large to fit our souls, and our souls are much too weak to guide technology's evolution (Bergson 1932, p. 267). This gap between our extended organs and our limited mental and moral capacities already led to enormous international political and socioeconomic issues in Bergson's time.

Technologies are extensions of biology. Where most animals are restricted by their species-specific physique as it evolved, some are able to use objects in their environment in such a way that they form extended organs. Beavers build dams that change the waterscape surrounding their burrows, which is functional for their survival, as well as for a host of other species (fish, bears, birds etc.). Some crows, a very intelligent bird species, now know when the traffic lights at a crossing turn on and off, and use them to safely place walnuts on the street surface, so that the wheels of the cars passing by can crack the hard shells, whilst they will pick up the soft interior pieces only when the light turns red again. But no species has evolved to use and modify objects from nature to such a creative manner as we do. In human technological history, a vital impulse fostering creative evolution clearly seems at work.

A general tendency of the life sciences is to understand the organism on the basis of the model of the automaton (Wong 2014): understanding life on the basis of the machine model, evolution and metabolism on the basis of a clockwork mechanism, and even the neurological functioning of the brain is described in mechanistic terms. Vitalism turns this around, opting for a reversed perspective, by studying machines as organic entities, as prostheses, as extensions of an organism. In other words, technology is considered from an organological perspective. if we follow the idea of organology, and consider our organs – our arms, legs, teeth, bowels, eyes, and ears – as naturally evolved instruments, we may reversely consider our instruments as artificial organs. This is the basic view Bergson takes on technology. It would be further developed by philosophers such as Georges Canguilhem, Gilbert Simondon and Bernard Stiegler, whilst traces of this view can also be found in the works of the palaeontologist Pierre Teilhard de Chardin and the archaeologist André

Leroi-Gourhan. The role of intelligence is to fabricate such external organs and guide the actions of our bodies to the bodies and objects in our environment. Science has greatly extended our technological potential for extension. Moreover, one could argue that even its direction seems to change. Whereas traditional technology is aimed at strengthening our control of the external world, new technologies are now entering our own bodies and brains, for instance in the context of neuro-enhancement, thus opening up a new chapter in the creative evolution of technology (Zwart 2017).

The basic property of technological artefacts is that they are undergoing a continuous process of refinement. In this regard some might speak of a targeted evolution, beyond the confines of Darwinian variation and selection. But to many engineers, it is basically a combination of innovation and serendipity: a form of accelerating evolution. In other words, the more or less spontaneous evolution of technology appears to be even more efficient and innovative than targeted and pre-planned evolution can be. It is the unexpected, the unplanned which generates more results than types of research based on blueprints for the future. Again, to understand technology from a Bergsonian perspective, we should not see it as a pre-planned application of mechanical reason, but rather as a process of creative evolution.

The German philosopher Peter Sloterdijk believes we are at a stage in which we are forced to change our lives: in *Du mußt dein Leben ändern* (*You must change your life* (Sloterdijk 2009), named after the last line of the famous Rilke poem *Archaïscher Torso Apollos*), he defends the idea that we should start applying 'techniques of individual and collective self-transformation' through which we can learn to restrict ourselves. At present, the whole organo-technical complex – our extended body – appears to have become auto-cannibalistic. It has assumed a tendency to dictate new needs to humans, without bothering to fulfil their existing needs, regardless of whether they are material or spiritual in nature. The gap between the limited capacities of our psyche and increasing powers of our organo-technically extended bodies has only grown immensely during the past century. It is now affecting the health of the planet and of the humans who inhabit it. Canguilhem adheres to the philosophical conviction that our mechanistic systems and their roles and representations should not be set apart from life. In his view, they rather form an integral part of living systems. Mechanics and machines are, in this view, modalities of the organic world (Wong 2014), rather than being separate from them. But the systems we created to carry our societies appear to have become destructive and entropic, rather than negentropic or self-organising. So, if we define life as autopoietic, self-creative and self-organising in nature, the human life form, under the sway of technology, has diverted from this path. Possibly, humans and their technologies should still be seen as modalities of the organic world, but if this is the case, it has grown into a rather problematic, highly disruptive modality. Still, this may not inevitably be the situation in which we find ourselves, provided that we take responsibility over our collective actions. It is here that the political and cultural dimension of Bergson's philosophy becomes relevant.

7.2   The Autopoietic Nature of Technological Systems

Arthur Schopenhauer, as should be clear from my earlier citation of the dark and pessimistic fragment, put human existence in a galactic perspective: space is endless, time has beginning nor end. We are merely a product of the mouldy film that covers the cooled down crust of one of many globes. As said, Friedrich Nietzsche alluded to this same image. We are, in Schopenhauer's and in Nietzsche's words, a being that, although it has the ability for knowledge, will hardly impact the order of the universe of we look at it on a grander scale, while time is endless and other processes have a much more enduring influence than our petty pursuits that we naively believe to be central to the whole of creation. In short, tan unjustified form of anthropocentrism remains in the Anthropocene-concept. At the same time, given the fact that we have the capacity to *know* what we are doing, we should nonetheless feel responsible for our current impact on planet Earth.

Over the past two centuries, our tendency to reproduce spurred an enormous growth of the human population. This, combined with a global increase in welfare standards and the level of consumption of goods, became an important cause of global warming, but it also resulted in large-scale migration, intensification of agriculture worldwide, and, as a result, the destruction of natural ecosystems. And our political discourses, once based on the self-evident value of democracy, are shifting towards large-scale manipulation of the masses through social media, thus enabling new forms of leadership of an oligarchical and sometimes even autocratic nature. Current developments in artificial intelligence and robotics shift our notions of personhood, the nature of work and employment and the very structure of our societies. There is also concern for the progressive digitalisation of our society: there is concern for the 'singularity' (coined by von Neumann in the 1950s, elaborated by Vinge (1993) and Kurzweil (2005)), the point at which artificial intelligence becomes smarter and more powerful than the naturally evolved intelligence of human beings.

These transitions affect the human condition on a very profound level. They are intrinsically connected to innovations (robotics, the internet, social media, genetic modification, etc.). But they appear to be the symptom of our Zeitgeist as well. Science and technology yield many useful products, but they currently seem to be, in the words of Henri Bergson (1932) 'expansive', without being 'progressive': an existential regression posing as technological progression. We always fail to grasp the deeper undercurrent of our own time.

The traditional ambition of science was to establish a fixed and valid representation of the world, but philosophical debate concerning the exact sciences has now resulted in the acceptance of David Hume's intuition that our scientific knowledge of the world is fundamentally uncertain (Prigogine and Stengers 1997). The aftermath of this blow to the traditional ambitions of science can still be felt, although many branches in the exact sciences are still in denial: the ideal of universal validity still guides many of the underlying assumptions of contemporary research in areas such as molecular life sciences, nanosciences and neurosciences. However, the blow also opened up new avenues for alternative approaches to scientific research.

Notably complexity theory became an impulse for novel ways of thinking about reality.

Bergson considered technology as a manifestation of the vital impulse outside of, or beyond biological evolution. As was mentioned earlier in this chapter, he considered technology as individuation (Bergson 1988 [1896]; Banerji 2015), by which he referred to a process of creative differentiation. The advantage of such a reading of technology is that it circumvents the mistake of seeing humanity as the sole author of technology, as captain at the helm of the ship of Progress, navigating through the unchartered seas of innovation, able to amend processes of diversion generated by technology. It also avoids a view on technology as merely determined by internal mechanisms, following a predetermined path, a view which amounts to ontological determinism applied to the level of technology.

The notion of complexity is not easy to explain and even more difficult to work through. One important property of complex systems is that processes in complex systems are irreversible (Cath 2018). They are non-linear and non-determined. Complex systems are auto-poetic in nature. They demonstrate a dynamics of self-emergent properties, within which innovative structures emerge, resulting in individuation in undeterminable ways. The internal interactions of a complex system cannot be fully explained on the basis external pressures. These interactions follow paths of convergence and divergence. This also means that there is no possibility for knowledge that can fully determine the system.

As discussed in Chap. 3, in complexity theory, two schools of thought persist: the one claims that although theoretically possible, it is practically impossible to reduce complex systems to the interactions of its components. The other claims that complex systems are not only practically speaking unpredictable. I will refer to these as the practical and the fundamental view.

The practical view embraces epistemic pragmatism: the view that although complex systems might hypothetically speaking be reduced to their causes, it is not possible within the grasp of human knowledge systems to know all inputs into the system. For those adhering to this view, complex systems theory offers an alternative to be able to still make useful predictive claims concerning the behaviour of such systems. But essentially, complexity is merely a high degree of complicatedness. If we order a bouillabaisse, this view entails that although nobody can revive the fish, onions and garlic, nor realign the individual atoms of the salt grains that went into the dish, such a reversal of processes is not impossible in the absolute sense of the word.[7] It is merely beyond the power of the human being. Here, complexity theory is merely a methodological tool, whilst the essential ontology underlying the worldview remains one of linear determinism. And essentially, reality can still be modelled.

The fundamental view entails a fundamental criticism of the idea that material determinism, be it linear (monocausal materialism) or non(or multi)linear (multicausal materialism) can explain every phenomenon in the universe. It is the latter

---

[7]And if the universe is infinite, it becomes quite likely there is a place where we can witness such processes.

version that is the focus here. It implies an alternative take on the role of time in the dynamics we experience and observe in the world. It dismisses the validity of mechanicism as such, criticising its continuing denial of the temporal nature of the universe, our experience of that universe and our observation of that universe. This entails a radically different way of thinking: it is fundamentally impossible to reduce the fish soup to its ingredients, since this negates the complex processes of change that went into its making, the many unknown unknowns that have contributed to its taste.

The sweat under the armpits of the cook set on by his hangover, caused by his neighbour's inferior pastis; the scent of roses surrounding the restaurant's entrance; the lipstick traces on the teeth of one's dinner partner; the panic of the fish when it was lifted out of its salty habitat in the sea; the cloud that briefly shadowed the sunlight when the waitress put the bowls on the table; the texture of the paper napkins under the cutlery; the proximity of the restaurant's restrooms to one's table; the potted mother-in-law's tongues on the restaurant's window sill, and so on: the question is one of epistemological expectation management: can one map these unknown unknowns, or are the processes in which they influence the experience of tasting the bouillabaisse fundamentally unpredictable? This depends practically on which mode of knowledge generation we functionalise, and fundamentally on which ontology of the universe we adhere to: a distinction related to one's aim and one's belief systems.

Many scientists are epistemological pragmatists, and adhere to the practical view: the idea that complexity can in theory be reduced to complicatedness. In other words, they hold the view that although it is practically more sensible to look at the infrastructural traits of complex systems, in principle they can be reduced to linear causal relations. Material determinism can still be reconciled with complex systems theory. There are however reasons to draw this reconciliation into question.

Material determinism still builds on a view on matter that predates the emergence of process thinking. It still takes material particles as the basis of any system, and physical forces as explanatory principles for their interactions. Complexity theory, however, has its own proper explanatory logic and elegance. Its use of emergence and self-organisation as explanatory tools should be viewed as an alternative to the material deterministic worldview of mechanistic thought. The behaviour of complex systems cannot be explained without a view on process thinking. In contrast to material determinism, process thinking departs from the view that the unfolding of processes is more fundamental to our understanding of reality than the analysis into the elements from which they are made up. Process thinking does not isolate things outside of time, but considers them as co-constituting time as such. It therefore carries along an alternative view on the concept of time. Bergson's philosophical vitalism does not propose an additional regulative force to explain the irreversibility of the arrow of time in organic life (as opposed to inorganic physical processes that appear to be reversible). With his vital impulse he rather proposes a description of self-organisation, by implementing concrete time (duration) as a factor.

The philosopher of science (and former physics and chemistry teacher) Gaston Bachelard coined the concept of an epistemological obstacle (Bachelard 1986 [1938]). With this term, he refers to tenacious and prejudiced ways of thinking that hamper the development of a scientific worldview. Thinking nature in terms of processes is only possible if scientists are able to overcome the epistemological obstacles of substantivism and mechanicism: the idea that the universe consists of a collection of things (not merely tables, stones and cows, but rather atoms, or subatomic particles) that are positioned in different ways at different moments in time. If this were the case, every state in the past would be conditional and completely deterministic for every state in the future. The present would then be a completely arbitrary and infinitesimally small dot on the line that extends from the beginning of the universe to the end of time. But if all is present, and past and future nothing more than constructs of memory and imagination, change is the only feature we directly experience. To be able to grasp the nature of change means to embrace complexity as fundamental. But this means that we should emancipate ourselves from the prejudices that gave rise to early modern physics and opt for a more creative and dynamical view of nature.

It is in this context that Bergson's reversal of teleology (discussed in Chap. 2) makes sense. Material determinism still embraces the idea of a predictable end point of processes, even though the nature of such end points is no longer regarded as meaningful in any metaphysical sense. Complexity theory regards the 'telos' (goal) as the driver of change. Complex systems are highly dependent on the 'white noise' that nurtures creative processes, rather than forming a disturbance of predefined, goal-oriented processes.

In contemporary science, complexity theory continues to compete with material determinism. Where in the past, classical mechanics, or more broadly, classical physics defined the scientific worldview, in recent years the life sciences have become the master paradigm of the sciences. The ghost of Newton that haunted scientific self-understanding is now increasingly under pressure, giving way to the break of day of complexity thinking. Whilst physics has progressed far beyond the confines of classical mechanics, the classical framework it still quite dominant as a source of inspiration (but eventually as an obstacle) in other fields, specifically those in which science is wed with engineering: molecular biology, nanotechnology, synthetic biology. In the current shift from fundamental scientific research to an engineering perspective, this master paradigm is still influential in the emerging disciplines of biotechnology, genetic engineering or CRISPR-Cas. These disciplines build onto an analytic approach to science. They aim to understand certain aspects of life by first being able to break it apart into constituent parts, and then being able to recombine or recompose the structures involved. This engineering approach to life is often guided by metaphors related to machines (car engines, the chassis of a truck, Lego blocks, and, of course, the clockwork).

The shift away from theoretical physics as master paradigm to the life sciences therefore paradoxically seems to amount to a consolidation rather than a weakening of the mechanistic worldview. The ghost of classical mechanics persists, and continues to haunt other fields: the modern neurosciences, neuropsychology, psychiatry but also sociology. Modern approaches in physics have failed to inspire other sciences to the degree that Newtonian classical mechanics once did, although the impact of thermodynamics is visible in other fields as well, while early modern clock-works of Huygens and others have evolved into cybernetics. From a Bergsonian perspective, however, we have failed to use the full potential of these developments and their metaphysical implications to understand and foster the ongoing revolution in the sciences. The tendency to relapse into mechanicism is still present.

The life sciences affect evolutionary processes directly, through modification of life forms (e.g. genetically modified organisms) on the basis of the recently discovered molecular principles of life. The creative flow of evolution is no longer bound to Darwinian processes of mutation and recombination. But it seems that we only marginally possess authorship of this major shift in the evolution of organo-technical systems. Techno-evolution steers itself and we have only begun to understand the processes we are involved in, perhaps merely as vehicles.

Walter Benjamin[8] developed a diagnosis of his time as an era in terms of mechanical reproduction. As he phrased it: "The presence of the original is the prerequisite to the concept of authenticity. Chemical analyses [sic] of the patina of a bronze can help to establish this, as does the proof that a given manuscript of the Middle Ages stems from an archive of the fifteenth century. The whole sphere of authenticity is outside technical – and, of course, not only technical – reproducibility." (Benjamin 1968 [1935], p. 220). In this regard, we live in a world that continuously distances itself from the authentic, since it leads to a continuous abandoning of 'tradition' via technological renewal (for instance by studying or reproducing books or paintings via computer screens). Humanity thus finds itself in a state of constant renewal. And technology, whilst forming an integrative part of the human condition, holds an alienating function in its ability to reproduce. This disenchantment, whilst laudable from Benjamin's Marxist perspective, is the challenge for a moral philosophy fit for our age and our globalising societies, since it also damages creativity.

So often quoted out of its context that it has become canon, Ludwig Feuerbach seems to premeditate these conclusions of Benjamin: "But certainly for the present age, which prefers the sign to the thing signified, the copy to the original, representation to reality, the appearance to the essence… illusion only is sacred, truth profane. Nay, sacredness is held to be enhanced in proportion as truth decreases and illusion increases, so that the highest degree of illusion comes to be the highest degree of sacredness." (Feuerbach 1848). Feuerbach aimed to discuss how we have become estranged from the true nature of religion, but if lifted out of the context of his discussion of Christianity, his quote may be taken to describe contemporary culture, filled as it is with digital images that replace realities to such an extent that we are no longer able to discern between the real and the image.

---

[8] In terms of the similarities and differences between Benjamin and Bergson: he agrees with Bergson that the structure of memory should inform philosophy's account of experience. But Benjamin was also critical of Bergson's tendency to generalise duration beyond and above historical experience (see Benjamin 1977 [1939]).

Benjamin's ideas focus on how modern technologies of reproduction, as part of capitalism, introduced movement into images. Benjamin thus sketches a tension between the industrial function of 'mass reproduction' of what used to be unique and the temporal nature of these reproductions (Lazzarato 2007). Bergson however, in his last work (1932) emphasises the originally mystical nature of technology through the interconnection between humans and technology. As such, his philosophy aims to overcome the tendency towards a purely negative assessment of technology that hovers over Walter Benjamin's philosophy as well as his own earlier works.

Reproduction does not only relate to books that can be printed, or works of art that can be photographed. Due to advances in microbiology, it now also affects living systems. They can be mimicked ('biomimicry') or even directly reproduced and recreated: synthetic biology is the field that propagates such promises, and, in some cases, already claims to have fulfilled them. Bergson already premeditated this development: for him, technology supplements biology. Intelligence, to his mind is a 'manufacturing kind of thinking'. In this regard, technology and innovation produce new machines and new ideas, but also new ways of living (Bergson 1911 [1907], p. 137).

Bergson gives a biological explanation of 'the genesis of technology' (Wong 2014) in which mechanical invention is interpreted as a biological function that is part of evolutionary processes. Bergson does not pit technology against the 'natural' human state, but considers it as an inherent part of human evolution. In this respect, matter can form both an obstacle to freedom and a precondition to overcome our natural limitations (Bergson 1932, p. 94). As Canguilhem phrases it, Bergson's philosophy aimed "to understand the true relationship of organism and mechanism, to develop a biological philosophy of machinism, to conceive machines as the organs of life, and to lay down the base of a general organology." (Canguilhem 1947; Wong 2014).

We are increasingly transformed by the technological systems we develop, and our lives, our societal institutions/structures, our communications and our ways of identifying with the world are increasingly incorporated in such systems. Our connection to each other evolves into hyperconnectivity: we develop and display our identities more online than in real life, more in interaction with machines and devices, as well as mediated through so-called 'machine-to-machine communication' than in direct, unmediated interaction with our fellow humans and social and natural environments. Our social relations (as well as our relations to ourselves) are increasingly digitalised (Facebook, email, WhatsApp). Also, the boundaries between the natural and the digital and between the individual and the collective have become vague, and even our sensory perception of the world has become increasingly technologically mediated.

Contemporary biosciences combine properties of biological research and of engineering. In view of this, evolution has become intertwined with technology. But rather than subverting nature to culture, or evolution to mechanics, the artefacts that science produces are increasingly demonstrating evolutionary tendencies in themselves, albeit evolution of a higher aggregation level, which is no longer bound by the restrictions of natural variation through mutation and recombination in the chromosomal structures of living organisms. Similarly, there have been revolutionary developments in neuroscience, in digitalisation and informatics, in the development of nanotechnology etc. which from Bergson's perspective can be seen as creative evolution. Whilst these developments are already convergences of different disciplines, they are also increasingly intertwined amongst each other.

Schopenhauer's dark fragment at the beginning of this section lays the foundation of an alternative, external perspective on our environment: rather than being at home in our natural environment, it alludes to the indifference of the universe towards the vulnerability of our planet, and the living systems that emerged on its surface. At the same time, it created the possibility of regarding all living things on the earth's surface as one complex system. The same is possible for the complex of all knowledge and information, and the complex of all technology.

In 1875 the concept of a biosphere was coined by the geologist Eduard Seuss. He defined it as the place on the earth's surface on which life dwells (Seuss 1875). Whilst the geosphere refers to the smouldering ball that constitutes our planet, the biosphere is the frail layer of life that emerged on its encrusted surface. It is the collective systems of living organisms on the surface of our planet. In addition to this biosphere, a noosphere is now evolving. It is a term that was coined by Vladimir Vernadsky, adapted by Bergson's successor Édouard Le Roy, and made more broadly known by Teilhard de Chardin. Noosphere is the collective system or web of knowledge, politics, thinking and communication that including all knowledge on the internet (the term noosphere is derived from the Greek term νοῦς, the mind).[9]

Bergson notably inspired Teilhard de Chardin's views through the former's biological account of the origin of technology. Rather than contrasting technology with evolution, Bergson considered technology to be an intrinsic aspect of life. As Teilhard de Chardin articulates it in *La Vision du Passé* ('A Vision of the Past', Teilhard de Chardin 1966[1923]) and other writings, we are distinct from other animals due to our ability to create functions or instruments that are not embedded within our body. Where birds both *have* wings as instruments for flight and *are* instruments of flight, and ants instrumentalise different functions within their species (soldier ants, queen ants, worker ants etc.), mankind has the ability to create such functions through extensions outside of the body. Thus, the nature of technology lies in our ability to externalise organs. This externalisation applies to functions of our body – walking (driving), digging (shovelling) or biting (cutting) – but also to our consciousness – recalling and speaking (writing), visualising (drawing) and so on. These externalised functions are not wholly dependent of our intentions. Their complex interactions with us and with each other rather give rise to processes of independent auto-poesis or self-creation. Teilhard de Chardin uses the term 'cosmogenesis' to refer to the process of increasing complexity, self-organisation and

---

[9] To this we might add the term technosphere: the collective systems of manmade technological artefacts, including digital networks as they emerged from the last quarter of the twentieth century (from the Greek τέχνη; that what is made).

ultimately self-awareness of the cosmos: from inorganic structures, organic structures emerged. From these, life evolved. But the emergence of consciousness is of a different nature. Animals may be aware, may feel, but they do not know how they know or how they feel. With the emergence of hominids and ultimately homo sapiens from the animal kingdom, life, as it were, turns towards itself. Evolution becomes aware of itself.

Immediate access to sensual awareness exists on the level of animal instinct. But being conscious of such awareness automatically creates distance. It is our ability to recall past situations, and to imagine future ones that lies at the basis of human freedom. The inorganic world is more or less bound to the determinants of its current states, the organic world is less so. Animal life demonstrates the ability to function in a more or less creative fashion – not fully determined by external impulses and instincts. Human life, due to memory and imagination, acts on the basis of visions concerning past and future states, and is therefore even more free to develop in a creative fashion. Bergson associated freedom with the virtual nature of memory and imagination. Whilst we are necessarily bound by matter, the level of complexity of the human being, in terms of the complex interactions of our nervous systems, provides for an independence of matter. Now it might be tempting to perceive of this notion of complexity as mere complicatedness. In other words, if atoms and molecules are completely determined in their behaviour, then, essentially, higher aggregation levels of what is ultimately still interactions between atoms and molecules should still be completely determined, although it may be nigh impossible to map how. But this would only be the case if we still retain the classical linear, constant time. The notion of complexity carries in another view on time, as creative evolution.

As Ansell-Pearson describes it, *Matter and Memory* is about a "movement from matter to memory" (Ansell-Pearson 2005). This movement is at the same time a movement from matter to freedom. As Ansell-Pearson phrases it: "no living body exists as a mathematical point in space, and no perception takes place in a mathematical instant in time" (Ansell-Pearson 2005). Deleuze followed this specific strand in Bergsonian thought in his treatment of the virtual. Time should therefore not be seen as something 'interior in us' (psychological, subjectively experienced time, which was Einstein's interpretation of Bergson's notion of duration) – but, in Ansell-Pearson's words, as "the interiority in which we move, live and change". It is not a virtual construct, but the nature of the virtual itself: and as such, it constitutes human freedom, whilst the clock time of the more traditionally inclined realists truly is a virtual construct.

The emergence of consciousness forms a gigantic leap in the history of the cosmos, comparable with the emergence of life itself. As an animal, or better, a living being that possesses the capability for rational thought and language (ζῷον λόγος ἔχων), humankind transcends the limitations of natural evolution. We might not be unique in being gifted with language, but we are with the combination between language and technology. As indicated earlier, Edward Seuss came up with the concept of a biosphere (Seuss 1875): the sum of all ecosystems that spans the globe, enveloping the inorganic geosphere, while Édouard Le Roy, Bergson's successor at the Collège de France, coined the term noosphere: the sphere of thought, reflection and information that encapsulates the earth. It can be regarded as a modulation of the existing biosphere and as a product of creative evolution. The noosphere emerged as a consequence of an evolutionary increase in complexity and consciousness. The emergence of humanity coincides with the emergence of a noosphere. it emerges through the human, not so much as rational animal, but first and foremost as a technical animal.

According to Bergson we should see ourselves as 'Homo faber' rather than as 'Homo sapiens'; the labouring human rather than the wise human. And as such our existence is determined by our ability to invent technologies. Technologies however are a pharmakon, in the original meaning of the word: poison and medicine at the same time (see also footnote 6). Individuals can hardly take a more than symbolic role in the global gambit that innovation has created regarding pollution, overpopulation, unequal distribution of wealth, knowledge and technology, protection of nature and biodiversity, exhaustion of natural resources and the limitations of planet earth. At the same time, we can no longer permit ourselves to believe that we are a comfortable part of nature. Our global impact calls for a completely new understanding of responsibility.

Bergson's (1911 [1907]) formulation of 'Homo faber' runs counter to traditional understanding of humans as thinking, knowing creatures. All social phenomena are part of the noosphere which is constituted by the interactions of humans with social phenomena. They include legal, educational, religious, research, industrial and technological systems. In this sense, the noosphere is an extension of the biosphere. As such, "social phenomena are the culmination of and not the attenuation of the biological phenomenon." (Teilhard de Chardin (1966[1923]), pp. 71, 230 and 261].

Science and technology should be viewed as a natural aspect of the human life-form rather than as an artificial attribute. And in this context, it is technicity rather than knowledge that typifies mankind. The aspects of labour and technical skills are intimately related to our being-in-the-world. The shift in the life sciences, from a vocabulary of discovery to a vocabulary of invention, holds an unexpected side effect. It shifts the distinction between epistemological subject and object: whilst homo sapiens might be interpreted as merely a side-branch of the Hominoids, it is at the same time different from all other species because it has evolved the ability to create its own functions outside of the body: bow and sling (arm), knife and arrow (teeth), shovel (hand), hammer (fist), house (hole), sewerage (bowels), car (legs), roads (soles), telephone (voice and ears), computer (brain), power grid (nerves) etc. This puts humanity in a fundamentally different category from other species, in spite of genetic or phenotypic semblances (Teilhard de Chardin 1966[1923]). We have now arrived at a point in history where our technological abilities not only create prosthetic organs outside of our bodies, but inside the body itself. Although artificial teeth have been around since the Etruscans (who fashioned them from human and animal teeth) true changes to the human physique and genetic traits were not within our grasp until recently. And although our brains have coevolved with the emergence of technology itself (a paleontological phenomenon lucidly dubbed 'the lithogenesis of mankind' by Peter Sloterdijk[10]), direct technological interventions in human nature were not possible until the advent of modern genomics and modern biotechnology.

Whilst Bergson refers to the evolution of technologies as fostering 'individua-tion', Gaston Bachelard, building on Carl Gustav Jung's theory of the development of consciousness, frames individuation in relation to the emergence of modern tech-nosciences. With Bachelard, Bergson shares the emphasis on the disclosing, world-creating aspect of technology, but as a process thinker, he perceives of the individuation of technologies as an open-ended (negentropic, autopoietic, self-organising) process. At least, this is the positive potential of technology, but unfor-tunately this is not always its actual manifestation.

It is the degree of complexity that creates a difference between the three layers of the abiotic, the living organic and the mechano-technical. The abiotic world already displays emergent properties in the process of (self)organisation, although much less complex and creative, and therefore, to Bergson's mind, it is not 'living' in the true sense of the word. The step towards the organic world signifies a higher level of organisational development. As a next step, the advent of humankind signifies a similar higher-order organisational process, comparable to the distinction between inorganic and organic, living and non-living matter. Where Darwin's theory of evo-lution applies to the organic world of plant and animal life, the human is a being that

---

[10] Lithogenesis: birth from stone, in reference to the parallel evolution of complexity of stone arte-facts and the size of the content of our skulls.

relies on the exploitation of technology. Technology goes beyond the constraints of 'natural' evolution, for instance in terms of directedness and pace, although ultimately it is based on the same dynamics as natural evolution. Human creativity, as expressed in our intimate relation to technique and technology, thus demonstrates the same creative evolutionary principles as biology as such.

Any view on technology needs to encompass a view on the synergies between human awareness, technology and nature: it needs to explore the nature of this synergy. Without the capacity of self-reflexive normative agency, we would find ourselves on that ship of fools. Our world would be reduced to collective behavioural tendencies, destined to disappear. Our very ability to reflect (self-conscious evolution) would be a fleeting moment, and deservedly so. Or is the very idea of Homo sapiens an illusion? The auto-poetic nature of technological systems evolution entails challenges for its generator. Bergson's aim is to develop a morality that transcends the limitations of different nations, ethnicities, and cultures. In this regard, the potential of technology to achieve true freedom – which he still describes positively in *Creative Evolution* (1911 [1907]) – has a darker aspect: technological progress is not incremental, a step forwards, but – as mentioned – it is expansive, a step outwards. It does not achieve true refinement, nor spiritual elevation by itself. It primarily achieves inflation. And in this regard, it may even assume a destructive guise, destroying ecosystems, poisoning natural and human habitats, digitally conditioning humans into a semi-vegetative existence predominantly dependent of virtual forms of life. We can only achieve a step beyond the destructive aspects of technology if we embrace a global policy based on appropriate insight in emerging solutions and on acknowledging complexity (Cath 2018). In other words, our ethics and policies should reach the same level of complexity. Here again, we should consciously rely on vital impulse, also when it comes to developing a global morality, to come to grips with the Anthropocene.

Bergson was not the first to observe the intertwinement of human nature with technology. It is even a key motif in Greek mythology. Notably the myth of Prometheus gives an account of the genesis of humanity with an explicit reference to technology. In the first volume of his series *Technics and Time,* entitled *The fault of Epimetheus* (1998), the French philosopher of technology Bernhard Stiegler takes this myth as a central line in his views on technology and its interrelation with the human subject. As Bergson said, myths have the ability to connect different generations and communities through fabulation – through what Bergson called the 'fonction fabulatrice': the myth-making faculty that we use to create persons and persona's to tell ourselves about ourselves (Bergson 1932). For Bergson, fabulation might save humanity from the vast, quick and disruptive technological developments of mechanistic modernity.

7.3   The Titanomachy Continued: From a Closed to an Open Morality

In her novel Atlas Shrugged (1957) Ayn Rand, the (in)famous Russian-American philosopher, also refers to the myth of Prometheus. She wrote that in order to take our destiny into our own hands, we need to become like the Titan Prometheus (Rand 1957). In order to become the creative beings that we are, we need to rid ourselves from the moral codes of self-denying altruism that govern society and condemn genius. Rand is known for her definition of egotism as a theoretical and pragmatic basis for an ethics for (ultra)libertarian capitalist societies. Disciples of her 'objectivist' movement include Allan Greenspan – economist and former chairman of the US Federal Reserve, Gene Rodenberry – screenwriter and producer of the Star Trek Franchise and Jimmy Wales – internet entrepreneur and co-founder of Wikipedia. In Rand's view, compassion and solidarity weaken the human. Once, she even expressed admiration for child rapist and killer Edward Hickman due to his embrace of his own nature without reserve – only criticising him for having been caught. But Rand's account of Prometheus is one-sided. The story runs deeper.

In Greek mythology, the Titanomachy is the battle between the old Titanic gods of the elements, and the new Olympic gods. The conflict represents a specific phase in the development of western societies. Ancient pre-classical Greek societies, in their religious systems, would focus on being dependent of natural circumstances, the elements, availability of resources, etc. The Titanic gods of the elements stood for water, fire, the earth, etc. The religion of the later classical Greek societies focused on different aspects of humankind: the organised society (Zeus) wisdom (Athena), the home (Hera), invention (Hephaestos) etc. In this development, one can recognize a specific shift in the human self-image, from a being that could not influence its environment to any large extent towards a being that became enabled to organise its environment. As such, this development also marks, in the words of Bergson, as elaborated by Karl Popper (1966 [1945]), the transition from a closed to an open society (Bergson 1932). It outlines the shift from being determined by natural circumstances to a relative autonomy from these circumstances. As such, it may be of interest to focus on one specific aspect in Greek mythology that symbolises this shift from proto-Greek society to classical Greek society.

After the Titanomachy, Zeus supreme sent most of the Titans to Tartarus to punish them for their revolt. But since the Titans Prometheus and Epimetheus had not sided with their aunts, uncles, and brother Atlas but with Zeus, the latter spared the two brothers. Instead, Zeus assigned Prometheus the task of shaping all creatures from clay and providing them with their δυνάμεις (dynameis), their abilities and capacities. Epimetheus implored his brother to be permitted to conduct this task. But whilst Prometheus had the gift of seeing, thinking and knowing ahead, Epimetheus was only capable of thinking, seeing and knowing in hindsight, by trial and error.

Epimetheus set about providing all creatures with the abilities awarded to them by the gods: the cat would have pointy ears, to better hear where the mice were hiding; the mouse was given the ability to dig small holes to hide for the cat; cows were given four stomachs to better digest grass and grasshoppers elongated hind legs in order to jump away from the cow's hooves. And thus, all of creation would be in balance. But when forgetful Epimetheus arrived at the last creature, he had already given away all properties. As a result of his mistake, this poor and unfortunate creature was born without sharp teeth to digest its food, without strong nails to dig itself a protective lair and without fur to protect it against the cold winter nights.

Prometheus, feeling responsible for his forgetful brother's mistake, took pity on this creature without properties, so he stole fire from Zeus' lightning bolt, as well as wisdom from his daughter Athena. He concealed it in a hollow stalk of fennel, and brought it to man. Thus, humans could warm themselves, tender their food by cooking it and scare off wild animals during the night. And in contrast with all other creatures, fire enlightened the mind: humankind could now make prosthetics for the lack of properties: knives instead of sharp teeth, bows and arrows instead of elongated hind legs, and axes to cut down the branches and trees with which to build a shelter.

Zeus became angered at Epimetheus's mistake and Prometheus's theft. He therefore tricked Epimetheus, asking him what he would desire in payment for his work. Epimetheus, ignoring his brother's warnings, asked for a companion to share his life with, and thus Zeus sent him the woman Pandora – "all-gifts" – as a bride. Prometheus crafted males, but the woman was created in the forge of Hephaestus: beautiful as a goddess.

Pandora was given a bridal chest by the gods, a box that she was instructed to keep closed. Prometheus, expecting retribution for his audacity, warned his brother against accepting gifts from Zeus. But Epimetheus was dazzled by Pandora's beauty and forgot the advice of his prescient brother. One day while her husband Epimetheus was away, Pandora opened the box Zeus had given them along with the warning never to open it. When she opened it, Pandora unleashed all the evils now known to humankind: war, plagues, and natural disasters. No longer could humanity loll about all day. From now on, we would have to work and would succumb to war, disasters and illnesses.

Prometheus was also punished: Zeus chained him to a rock in the Caucasus, and his liver was eaten daily by an eagle, only to be regenerated by night, due to his immortality. But in spite of Zeus's terrible wrath, he was not without mercy. The box was not left completely empty. One thing remained: elpis, both the hope and the fear of our anticipation of the future.

The myth of Prometheus and Epimetheus is described in very different forms by Hesiod, Aeschylus and Plato. About ten further Greek and Roman authors retold and further embellished the Prometheus myth from the fifth century BC (Diodorus, Herodotus) up to the fourth century AD. Sappho, Aesop and Ovid added the motif that Prometheus also created the human race out of clay. The story of Prometheus is a typical origin myth. It recounts the original technical nature of humankind. As such, it may aid in a self-analysis of human identity.

The above rendition is a collage from different sources, but the story as such spurred a philosophical interpretation by Bernard Stiegler (Stiegler 1998; Franssen

2017; Lemmens and Hui 2017). It is not merely the myth of the creation of human-kind, but also of humanity's intrinsic creative and technological nature. The motif of Prometheus creating humankind out of clay is likely inspired by the way in which humans created figurines out of clay himself. Since we have started to (re)produce ourselves and our social systems in the form of technological artifacts, we have become Promethean ourselves. This motif can be found in ancient Egyptian mythology as well. The myth of Prometheus thematises the step from a society that largely needed to focus on survival in nature to a society that largely needed to survive in an intercultural multistate context. It also signifies the emergence of human agency: the gift of Prometheus is the enlightenment of the mind, its curse is the necessity of responsibility.

In Ayn Rand's evolutionary moral philosophy, she equated 'survival' with hedonism on the one hand and moral justification on the other. But she excluded the place of agency, the thing that places us beyond traditional interpretations of the evolutionary principle. This is reflected in the deep significance of the Titanomachy. The Titanomachy, the struggle between the old Gods and the new Olympic Gods was won by the latter, and the only two survivors that were not punished by Zeus were Prometheus (foresight, imagination, negligent of the present) and Epimetheus (hindsight, memory, negligent of the present) as we have seen. But Prometheus betrays Zeus and the other Olympic Gods by stealing the fire of knowledge and giving it to the one creature that his brother, Epimetheus failed to sufficiently equip to survive in the world (in Darwinian terms: an un-fit animal). Ultimately, the Titanomachy was about the extent to which humankind should be entrusted with god-like responsibilities. And as such, the struggle is still ongoing. We find ourselves at a time in history when survival demands again surpass our given limitations. If we are not up to the task, we will make the planet uninhabitable for many species including our own. We cannot reside in the luxury of thinking ourselves as merely being amongst other beings. We cannot rely on evolutionary polishing as if we did not surpass the boundaries of DNA-bound evolution. We have postponed this responsibility for too long already.

According to Bernard Stiegler (1998) it is the forgetfulness of Epimetheus (Prometheus' brother) that caused our essential lack: an original incompleteness that always has to be supplemented by the technical. For Stiegler, this supplement is fundamental to being human. This supplement is not something additional, it is the very condition of mankind. Technicity is human nature, without it, we would merely be a naked ape, and likely would have gone extinct early on in our branching off from other large apes. Bernard Stiegler refers to the myth of Prometheus, and specifically the role of his brother Epimetheus (who gifted all creatures with their properties, but neglected humankind), in his account of our original fault. Technicity has coevolved with an original lack of the right organic properties to survive. Technicity does not merely produce artefacts. These function as prosthetics to supplement our lack of original properties. We produce technology, but technicity produces humanity as well. Since we have started to (re)produce ourselves and our social systems via technological artefacts, we are not merely Prometheus' creation; we have become Prometheus ourselves. Bergson's ideas about technology present humanity as a 'natural cyborg' (Marrati 2010): our abilities to create artefacts is at the same time our potential to recreate and reconfigure ourselves and our environment.

We have to manage in the world by creating our own organs. This defines the true nature of technology as well as the true nature of the human organism. As such, we are the product of a divine battle and the only escape is to fulfill our deepest ambition and to become god-like ourselves. Whilst most religions impose a ban on any attempts to become equal to God or the gods (the tower of Babel, the notion of hubris, etc.) the myth of Prometheus does suggest the existence of something like a divine spark in human being. We are not only toiling animals (subject to the power of Mount Olympus) but also knowing animals. We need to accept Prometheus' heritage and assume an active role in the current version of the Gigantomachy, because Anthropocene basically means that we have evolved into a titanic force ourselves.

Although we continuously come to know more about ourselves, through physiology, neurology, psychology and so forth, there is something about ourselves we fail to grasp in full and which nonetheless continuously reveals itself to us. This continuous revelation creates a need for new concepts, new vocabularies and new ways of thinking: a continuous process of 'reconstructing the self' (Landeweerd 2009) through self-images. It is here, in these efforts of self-portrayal, as conscious beings, that we continuously transcend the limitations of our biologically given nature.

We cannot begin to understand our position in this world without involving scientific knowledge about ourselves. To understand the nature of the human being necessitates an account of our inherently technological relation towards ourselves and our environments. Due to our lack of natural organic capacities, we are *Homo faber*, the labouring human being that needs to produce its own supplements via inventive technologies. At first glance, technologies seem mere instruments: we invent them to serve our purposes. But the evolution of technological systems can better be typified as a life form in its own right. We are not the sole authors of the technological systems and environments that determine the human condition. Like the evolution of species, the evolution of technological systems is a creative process[11]: it is not determined by mechanical parameters on the basis of previous states alone. But this does not mean that we are necessarily instrumental features of that evolution.

In *The two sources of Morality and Religion* (1932), Bergson embraces a view on technology as mystic in nature rather than merely being the mundane result of human invention. Tools, artefacts and machines have a deep relation to the human soul. We are typified by technologies as much as they are typified by us. He does so by reference to the human need to self-transcend. This is not possible merely on the basis of some natural ability, and not even on the basis of our capacity to speak. Bergson explains that "[…] if our organs [– our limbs, our intestines, our brain –] are natural instruments, then our instruments must be considered as artificial organs" (Bergson 1935 [1932], pp. 268). This insight, which reverberates with Heidegger's views on tools, Teilhard's understanding of the noosphere Stiegler's concept of a

---

[11] And, to reiterate the societal and religious stake of the discussion at the time and before that time, not a product of preset creation.

general organology, has enormous consequences for our conception of technology. Any mechanism is an extension and an exteriorisation of our body, granting us with abilities that transcend those of the bodies we are naturally endowed with, but also reflecting back on our nature by transforming *us*. However, in the past two centuries, the machines and mechanisms that extend our natural abilities and form a basis for our impulse towards transcendence have grown out of proportion (Bergson 1932), a theme that was later picked up by Simondon and others.

We neglect our global responsibility for the technologically mediated environments we inhabit. Next to intensified agriculture, our current use of fossil resources forms a major cause of global warming. The exploitation of fossil fuels for industrial purposes has spurred an unprecedented growth and gave rise to our permanent-growth-oriented economies during the nineteenth and the twentieth century. Without fossil fuels, the industrial revolution would have been unthinkable, or would have come to a halt quite early. Their exploitation gave rise to the introduction of trains, steamships and the industrial production of food, pottery, cloths, fabrics and multiple other products. Further accelerations were caused by the exploitation of oil (during the 1920s), the mass introduction of cars, the development of plastics and their large-scale production and consumption. The plastic revolution starting with Leo Henricus Arthur Baekeland's retail of Bakelite and culminated in the current global disaster with huge quantities of plastic: plastic islands the size of complete provinces, floating in the oceans (the Great Pacific Garbage Patch is only a first example). A further acceleration was initiated by the rapid growth of population, economy, use of natural resources, and communication after WWII. This post-war era explicitly entailed the introduction of economic models focused on the creation of needs, rather than on the production of goods to meet existing needs.

We find ourselves in a moment in history that is characterised by massive changes in the earth's biosphere. In the past decades, it has become clear that we do have a fundamental and enduring impact on processes that, until now, seemed hardly noticeable, let alone affected by us. As indicated, the influence of humankind on the terrestrial environment is of such magnitude that we have entered a new geological era: the Anthropocene. Natural fluctuation in climate pale in significance compared to our massive impact on planet earth since the beginning of the industrial revolution in terms of loss of biodiversity, climate change and pollution (Crutzen and Stoermer 2000; Crutzen 2002). In the scientific community, there is a clear consensus concerning the gravity and urgency of global climate change. The 'great acceleration', the rapid growth of population, economy, use of natural resources, communication etc. after WWII (McNeil 2014) has carried along grave impacts. The International Panel on Climate Change (IPCC) urges global societies to take action, but until now, socio-economic and political actors are either in denial (politicians such as Trump and Bolsonaro fall in this category) or slow to respond (most if not all others): too little is done too late. Humanity has a destructive influence on planetary life as climate change, loss of biodiversity, plastic pollution of the oceans and mass migration form a cluster of interrelated problems. The Anthropocene presents us with the necessity to step from static and relatively local forms of morality to an open and global morality. To develop such a morality, however, we have to start from the uniqueness of the current situation. Life as such already is unique, as we have seen because of its tendency to self-organise. Subsequently, one particular life form has developed the ability to produce tools and artifacts. The emergence of humankind goes hand in hand with the emergence of technicity. Although other animals can be mentioned that use artifacts as well,[12] the extent to which we humans coevolved with technology is unprecedented in the history of evolution. Technology is not only organised (by us, to some extent at least), but also self-organising. Modern technology generates new forms of complexity, but also entails disastrous, entropic effects. It is negentropic and entropic at the same time. It both increases and destroys complexity. It generates both order and chaos, it enhances and destroys vitality. Diversity and diversification are threatened on multiple levels, not only in the environmental real. The internet connects, but causes new frictions and collisions as well. Novel communication systems cause a dramatic decline in the diversity of languages, whether it concerns the language of academia and literature (English is becoming the standard virtually everywhere) or smaller local languages in Africa or South America. The imagination of both adults and children is reduced under the influence of addictive devices and artifacts that continuously compete for our attention. Research and innovation are tailored towards efficiency, production and, most importantly, speed. The 'great acceleration' shows no signs of coming to a halt, and this is putting severe pressure on human societies and natural ecosystems alike.

We do not fully grasp the nature of technology. Technology is craft, and as such it is a living relation to matter. Whilst we believe to be at the steer of its evolution, we are merely part of an intricate coevolution. To give an example: in Bonsai, the person pruning the trees might believe he is creating its shape. But to grow towards true mastership in bonsai, any pruner needs to acknowledge he is only following how the plant revealing itself. But this relation is not passive, it is, as Bonsai master Peter Chan also recounts (1989) creative. Similarly, we cannot determine the path of our coevolution with technology, but we are an active part of its evolution. We can only enjoy its creative nature of this relation if we remain open to the process of what is given, rather than deciding upon this. But in this age, we tend to neglect our ability to intuit the relation to life.

As Bergson remarked: "Human will not rise above earth if powerful tools are not provided to support. He will have to weigh over material in order to detach from it; in other words, the mystic calls for the mechanics; it has not been noted so far, because mechanics, by an accident of switch, was launched on a highway towards exaggerated well-being and luxury for a limited number instead of release for all" (Bergson 1935 [1932], p. 267). The mystic calling of humanity towards transcendence calls for new forms of mechanics. In other words, we must recover the mystical nature of technology, of mechanics, in order to take it beyond its current function, which is to generate luxuries for a limited number of people. Technology's true aim is to release the whole of humanity from its material confines. Our techno-evolution

---

[12] Beavers building dams, apes using sticks to angle for ants, parakeets using strips of paper to adorn their tail feathers, crows timing traffic lights to crack nuts under the tires of passing cars.

has so far remained a process which steers towards eco-systemic collapse, environmental pollution and human suffering rather than progress. Mechanics, technologies and tools have become dysfunctional rather than supportive to foster a genuine coevolution of the human being. They now rather pose a threat to the continued evolution of man. They are on a rampant course. However, this does not mean that this course is fixed and predetermined and that there is no room for creativity and responsibility at all.

The *Two Sources of Morality and Religion* would be Bergson's last major book. Although most authors writing on Bergson's philosophy prefer to understate the importance of this book (Deleuze for example completely ignored it in his *Bergsonism* (1966)), others, such as Vladimir Jankélévitch, considered it as a key publication. In *The Two Sources of Morality and Religion* Bergson introduces a distinction between static (closed) and dynamic (open) morality. Static morality is determined by its evolutionary function to protect the group and safeguard internal consistency. It is organised through religious systems, myths and rituals. It is a morality of obligation, functional for ordering communities of limited size, but at the same time encouraging a hostile attitude towards other groups. It is therefore a form of morality that predisposed to generate wars and conflicts between groups and communities. In the dynamic emergence of global networks, its evolutionary function has become counterproductive. The second, more dynamic form of morality focuses on a vision of humanity as a whole, in interrelation with mechanical and biological systems. This is the social value of mystic experience. This mystic outlook sees humanity as coevolving with technology. This is process with a mystical dimension because it entails the promise to release us from the confines of matter and take a creative leap towards a new, global human condition. Technology makes it possible to free ourselves from the confines of the past, so that potentially humanity become one with the divine. Technology is in essence mystical experience.

Closed morality is informed by instinct, by biological predisposition. It has an evolutionary function in protecting the group through group solidarity, against external threats. The static systems of religion that are informed by closed morality were once functional for a species whose individuals could not survive in a solitarily manner. Closed morality is an evolutionary phenomenon which evolved amongst social species. It is a type of morality that organises social structures, and is thus focused on survival. But such societies have become outdated. By inciting tensions between peoples and nations, they now unavoidably threaten human survival rather than safeguarding it. Although at one time functional for the survival of human beings, in a globalizing society it has become a threat rather than a force for the good. Closed societies embrace static religions that build upon the 'fabulation function' – our ability to create narratives that express a sense of a divine presence watching over us. It is our ability to invent images of gods. These images then require strict obedience to a closed morality. Although closed morality and static religion ensure social cohesion, their negative aspects (pitting one group with one static system against another) are now becoming painfully clear.

Similar to Popper's view on the evolution from closed to open societies, Bergson's sketch of closed societies (with their static moral and religious systems) comes dangerously close to a condemnation of supposedly primitive cultures, either of the past (Jones 2007), or existing in other parts of the world. The risk is that the sophisticated culture Western Europe of the 1930s is unquestioningly adopted as the leading criterion. Still, Bergson is convinced of the importance to develop a global (universal) moral culture. This explain why he became so strongly involved in the *Commission Internationale de Coopération* Intellectuelle (CICI – the International Committee on Intellectual Cooperation) having been appointed as its first chairman. This committee launched a book series under the title 'Internationalism' and aimed to establish a dialogue between different traditions across the globe. Bergson's notion of open morality does entail some important component for a philosophical diagnostic of the present situation.

Open morality and dynamic religion are concerned with creativity and progress. They are not restricted to specific communities or groups, but extend towards the global world, and invite exchange with what is different, other or opposite. Open morality and dynamic religion could, in Bergson's mind, only emerge in an open society (Bergson 1932). The notion of an open society was later elaborated by Karl Popper (1966 [1945]). Open societies are inclined to seek a universal morality, rather than a morality that is bound up with the specific identities of a particular cultural or ethnic group. They seek collaboration rather than conflict. They build upon imagination and creativity. Its religious systems are mystical in nature. Closed morality as a biological function cannot resolve the issue of human responsibility over the planet. It confronts us with the need to take a responsible approach to research and innovation rather than remaining within the confines of closed morality to protect the group. Emerging research fields have to find ways to better connect to effectuate awareness and change. There is an over-abundance of expertise when it comes to studying science and technology from societal, sociological, moral-philosophical and policy angles, but so far these research areas fail to connect to the growing demand for societal change. Our knowledge concerning auto-poetic evolutionary systems constitutes an unrealised potential. There insight may help us to act with wisdom. Although our role in the techno-evolutionary processes that are currently taking place can no longer be seen as leading, it cannot be dismissed as completely determined by external sociocultural, biological or evolutionary factors either. Research and reflection must engender new forms of collective agency and actions.

Like the emergence of languages and dialects, the complex processes of differentiation and convergence in the sciences can be regarded as life forms. They consist of intricate processes of individuation. The concept of individuation, mentioned in relation to Bachelard's works, was adopted from Carl Gustav Jung and refers to the process of becoming a complete individual, someone who managed to address the tensions between various dimensions of his or her personality, which are initially experienced as diverging. 'Individuation was also taken up by Gilbert Simondon (1924–1989), a philosopher of technology inspired by Bergson's notion of technology. In a broader sense, individuation, undividedness, can also be termed non-identification: the process of becoming non-identical to one particular aspect and, as a result, become 'all else'. This process of individuation can also be discerned in the history of languages and the history of science where, after periods of separation and differentiation (into branches, disciplines, schools, dialectics, etc.) we now see a process of convergence into a global language.

In his description of individuation, Simondon discussed Bergson's views of becoming, emergence and creative evolution in the context of cybernetics, chaos theory and complex systems theory. His philosophy encompasses an understanding of process of individuation in a broader than mere psychological sense: individuation is a actualisation of energy (or transduction) which can be discerned in physics (for example with regard to turbulence theory or quantum field theory), in chemistry (the study of liquid and aperiodic crystals); in psychology (new converging approaches to perception, affection and the unconscious), in mathematics (chaos theory and self-organisation) and in biology. In the latter field, research is now revealing how individuation emerged in the past via the incorporation of mitochondria in cells and of foreign or viral DNA is the genomes of plants and animals. Avoiding the teleological thesis, Simondon held that that the individual should be considered as an effect of individuation rather than a final cause of such a process.

Duration and élan vital are discernible in all living systems, and this goes for the systems of knowledge as much as for the growth of crystals or the behaviour of subatomic particles. But they can also be seen in the evolution of galaxies, the emergence of technological systems and the role of creativity in the arts. The difference between creative evolution in stellar constellation, life forms and technology is predominantly connected with the level of aggregation, the level of complexity. The emergence of new technologies is as much characterised by creative evolutionary processes as the emergence of new life forms as studied by biology.

Bergson distinguishes between material and spiritual energy. Spiritual energy emerged in the course of evolution as consciousness and self-consciousness. The modern sciences predominantly focus on energy of a physical (material) nature, although research priorities have gradually shifted from geometry to physics to chemistry and ultimately to biology (Bergson 1932, p. 313). The sciences managed to drill into fossil resources, organic and biological in origin, that had been stashed in the earth's crust for millions of years. In the form of coal, oil and gas, these energies were unlocked and suddenly were at our disposal. The availability of such vast amounts of energy is without precedent in history. From the late eighteenth century onwards, their application in transportation and production radically transformed our life world. Bergson believed that the discrepancy between the attention spent on material energy and the attention spent on spiritual energy led to a problematic situation. Infinite growth became a key principle for our modern industrial societies, but our capacity for critical reflection was neglected. As a result, the utilisation of fossil energy became exponentially expansive, while our ability to steer its direction did not keep pace. We see the neglect of spiritual nature also in the inherent tendency in modern science to explain the world in spatial terms and this also applies to the scientific understanding of the human mind, which is reduced to interactions between neurons, while the spiritual dimension is excluded from the vocabulary of science. Here, Bergson again insists on a temporal account of spiritual energy (Bergson 1932, pp. 313–314). Our brains are not to be regarded merely as organs that create mental images and store them away. Rather, human consciousness is a temporal phenomenon, a consciousness of past, present and future: it means being-in-time. The complex relationship between organism and consciousness, between body and spirit, is not fully captured science. This would require the elaboration of a comprehensive metaphysics (Bergson 1932, p. 316).

In short, Bergson's claim is that due to the focus of modern science on the material and the spatial, it has neglected the temporal and the spiritual. Is the temporal then only applicable to the mind (the soul) and is the material only spatial in nature? Throughout his oeuvre, Bergson insists it is not, although he does admit that duration is only applicable to that which can experience it, so that de facto it is restricted to the realm of life (although microorganisms, plants, animals and humans experience duration on different aggregation levels).[13] But in *The Two Sources of Morality and Religion*, Bergson does elaborate a temporal dimension for the physico-chemical realm. His account of the vast timespans during which oil, coal and gas and other fossil resources accumulated, seems to entail the notion of what is nowadays known as 'deep time'. The problem is, however, that humanity did not sufficiently develop an awareness of this deep temporal dimension. As a result, the advance of knowledge did not coevolve with a similar advance of our spiritual ability to grasp the ways in which these scientific insights affect society, ecology and the human psyche. In the end, Bergson urges to act rather than contemplate. Given the detrimental effect that our collective behaviours have on society, ecology and the mind, we need to rise above our tendency to consider the biosphere as an unchangeable given, independent from and resistant to human intervention. We need to consider this 'rising above ourselves' as a development that is part of our very evolutionary role: transcending the limited notion of ourselves as a species among species in and biologically given natural environment. Bergson describes this self-transcendence as the function of the universe: technology as a machine for the making of future humans as God-like creatures. Although this notion echoes the Nietzschean notion of a superhuman, Bergson does not want to imply that we are divine (neither did Nietzsche, for that matter). Rather, we need to rise above our biological preconditions and embrace the level responsibility belonging to a being that is capable of destroying its habitat, and as a result its own conditions to survive.

The evolutionary process of technology occurs through convergence and divergence as much as natural evolutionary processes do (Cath 2018). Technocracy presents predictability and control as means to achieve certain goals, but they are in effect the goals of technocracy. To be able to get there, technocratic management will force the dynamic and creative processes of scientific thought into the static and mechanistic molds. Thus, regulatory principles and protocols are given priority over the practices they are supposed to facilitate. Technoscience runs a serious risk of being instrumentalised to serve such technocratic managerial endeavours. But there is hope: technoscience might also come to terms with its own process-nature. It is, after all, evolutionary in nature. Some solace might thus be found in Bergson's view of the function of the universe as a machine for the making of Gods.

---

[13] Simondon was justified in his criticism of this limitation of Bergson's notion of duration; concerning Benjamin's critique, see footnote 107.

Humanity pays the price for the divine gift of knowledge. Although knowledge generates the tools that we need for our survival, when taken out of its pragmatic context it serves no purpose and becomes self-referential. Moreover, the ability to create tools for our survival has grown beyond proportion. The divine gift of knowledge demands that we surpass our evolutionary limitations – to consume, to protect the group, to fight the dangers of nature – that have, in the past few hundred years proven to be destructive for both our own species and the environments we inhabit. We have to transcend our biology and become divine in our own right to be able to survive and fulfil the essential function of this universe.

In Plato's Symposium, Diotima describes how the striving of mortals for immortality is expressed in their creative and reproductive tendencies. She refers to it in relation to ποίησις, an ancient Greek term that was derived from the verb ποιέω: "to create". This word, the root of the word "poetry", signifies free interpretative action on existing materials and traditions to continue the world through transformation. Achieving immortality, a continued existence of our species, means accepting change.

Literature

Ansell-Pearson, K. (2005). The reality of the virtual: Bergson and Deleuze. *MLN, 120*(5), 1112–1127, (16 pages), Comparative Literature Issue (Dec., 2005).

Bachelard, G. (1986 [1938]). *The formation of the scientific mind: A contribution to a psychoanalysis of objective knowledge*. Boston, MA: Beacon Press.

Banerji, D. (2015). Individuation, cosmogenesis and technology: Sri Aurobindo and Gilbert Simondon. *Integral Review, 11*, 1. See: http://integral-review.org/backissue/vol-11-no-1-feb-2015/.

Barry, A., & Maslin, M. (2016). The politics of the anthropocene: A dialogue. *Geo: Geography and Environment, 3*(2).

Benjamin, W. (1968 [1935]). Hannah Arendt (ed.). Illuminations [1940], Trans. Harry Zohn (London: Fontana, 1977); The work of art in the age of mechanical reproduction. In: *Illuminations*. London: Fontana; On some motifs on Baudelaire.

Bergson, H. (1988 [1896]). *Matter and memory*. Translated bys N. M. Paul & W. S. Palmer. New York: Zone Books.

Bergson, H. (1911 [1907]). *Creative evolution*. New York: Henry Holt and Company.

Bergson, H. (1927a). Banquet speech. In H. Frenz (Ed.), *Nobel Lectures, Literature 1901–1967*. Amsterdam: Elsevier Publishing Company.

Bergson, H. (1927c). Lettre à Léon Brunschvicg, 22 février 1927. *Journal des Débats, 28* février 1927.

Bergson, H. (1977 [1932]). *The Two Sources of Morality and Religion* (trans. Audra, R. A., Brereton, C.). Notre Dame: University of Notre Dame Press.

Canguilhem, G. (1947). Note sur la situation faite en France à la philosophie biologique. *Revue de Métaphysique et de Morale, 52*(3), 322–332.

Cath, A. (2018). *Solace in complexity, a fundamental empirical thought experiment*. Doctoral thesis. Vianen: Uitgeverij ProefschriftMaken.

Chabot, P. (2013). *The philosophy of Simondon: Between technology and individuation* (trans. Krefetz, A., Kirkpatrick, G.). London/New York: Bloomsbury Academic.

Chan, P. (1989). *A Guide to Bonsai*. London: Bracken Books.

Crutzen, P. J. (2002). Geology of mankind. *Nature*, 415–423.

Crutzen, P. J., & Stoermer, E. F. (2000). The Anthropocene. *Global Change Newsletter, 41*, 17.

Deleuze, G. (1991 [1966]). *Bergsonism* (trans. Tomlinson, H., Habberjam, B.). New York: Zone Books.

Derrida, J. (1981). *Dissemination* (trans. Johnson, B.). Chicago: University of Chicago Press.

Feuerbach, L. 1881 [1848 [1841]]). *Das Wesen des Christenthums*. 2nd edition, 1848 (online). (in English) *The Essence of Christianity* (1854) (trans. Evans, M.). St. Mary s. 2nd edition, 1881. Oxford.

Franssen, T. M. (2017). Prometheus Redivivus. The mythological roots of transhumanism. In E. Sampanikou (Ed.), *Audiovisual Posthumanism* (pp. 27–49). Cambridge: Cambridge Scholars Publishing.

Jankélévitch, V. (2008 [1931]). *Henri Bergson*. Paris: Presses universitaires de France.

Jones, D. (2007). Review: The Eleatic Bergson; reviewed work: Thinking in time: An introduction to Henri Bergson by Suzanne Guerlac. *Diacritics, 37*(1), 21–31.

Kant, I. (2007 [1786]). Conjectural beginning of human history. In *Anthropology, history, and education*. Cambridge University Press.

Kurzweil, R. (2005). *The singularity is near*. New York: Penguin Group.

Landeweerd, L. (2009). *Reconstructing the self: Issues of choice, fate and justification in the eugenics debate*. Maastricht: Datawyse.

Landeweerd, L. (2018). *Icarus' flight. Essay for the opening of the academic year at iArts Maastricht*. http://www.iartsmaastricht.com/uploads/images/2018-2019/public%20reading%20opening%20day%20iArts%202018%20(1).pdf

Lazzarato, M. (2007). Machines to crystallize time: Bergson. *Theory Culture & Society, 24*(6), 93–122.

Lemmens, P., & Hui, Y. (2017). Reframing the Technosphere: Peter Sloterdijk and Bernard Stiegler's Anthropotechnological diagnoses of the Anthropocene. *Krisis Journal for Contemporary Philosophy*.

Lewis, S. L., & Maslin, M. A. (2015a). A transparent framework for defining the Anthropocene epoch. *The Anthropocene Review, 2*(2), 128–146.

Lewis, S. L., & Maslin, M. A. (2015b). Defining the anthropocene. *Nature, 519*(7542), 171–180.

Marrati, P. (2010). The natural cyborg: The stakes of Bergson's philosophy of evolution. *The Southern Journal of Philosophy, 48*(2010), 15.

Mcneill, J. R. (2014). *The great acceleration: An environmental history of the Anthropocene since 1945*. Cambridge: Harvard University Press.

Nietzsche, F. (1999 [1873]). On truth and lies in a nonmoral sense. In *Philosophy and truth: Selections from Nietzsche's notebooks of the early 1870's* (trans. Breazeale, D.). New York: Amherst.

Nietzsche, F. (1968 [1885-1892]). Thus spoke Zarathustra. Walter Kaufmann (trans.). In *The portable Nietzsche*. New York: Viking Press.

Popper, K. R. (1966 [1945]). *The open society and its enemies, 2 vols* (5th ed.). London: Routledge & Kegan Paul Ltd.

Prigogine, I., & Stengers, I. (1997). *The end of certainty: Time, Chaos and the new Laws of nature*. New York: The Free Press.

Rand, A. (1992 [1957]). *Atlas shrugged* (35th anniversary ed.). New York: Dutton.

Schopenhauer, A. (1958 [1818]). *The world as will and representation volume I* (trans: Payne, E.F.J.). Mineola: Dover Publications.

Seuss, E. (1875). *Die Entstehung Der Alpen*. Vienna: W. Braunmuller.

Sloterdijk, P. (2013 [2009]). *You must change your life*. Malden, MA: Polity Press.

Stiegler, B. (1998). *Technics and time, 1: The fault of Epimetheus* (trans: Beardsworth, R., Collins, G.). Stanford: Stanford University Press.

Stiegler, B. (2012). Relational ecology and the digital pharmakon. *Culture Machine, 13*, 1–19.

Teilhard de Chardin, P. (1966 [1923]). The Vision of the Past (trans. J.M. Cohen). London: Collins.

Vinge, V. (1993). *The coming technological singularity: How to survive in the post human era.* https://edoras.sdsu.edu/~vinge/misc/singularity.html. Accessed 10 Oct 2019.

Wong, A. (2014). *Life and technology in Bergson's creative evolution.* http://hdl.handle.net/2268/186268. Accessed 13 Oct 2019.

Zwart, H. (2017). The oblique perspective: Philosophical diagnostics of contemporary life sciences research. *Life Sciences, Society and Policy, 13*, 4. https://doi.org/10.1186/s40504-017-0047-9.

Zwart, H., Landeweerd, L., & van Rooij, A. (2014). Adapt or perish? Assessing the recent shift in the European research funding arena from ELSA to RRI. *Life Sciences, Society and Policy, 10*, 11.

Chapter 8
Conclusion and Notes on Various Themes

This book was originally meant to become a chapter in an edited volume; its aim was to discuss the relevance of four neglected French philosophers for the philosophy of science – the others being Gaston Bachelard, Michel Serres and Gilbert Simondon. But during my preliminary rereading of Bergson's works, I became increasingly convinced that what I wanted to discuss could not be realised within the limitations of a book chapter. The subject matter I studied, as well as the inter-relation of the domains and concepts I encountered, demanded the writing of a book in its own right. It would take 3 years to achieve this. Another reason for writing this book was the fact that I noticed a critical attitude towards Bergson's oeuvre, among scholars working in philosophy of science and among exact scientists (if they were aware of the existence of Bergson's oeuvre at all). I hope this book will contribute, not only to a rehabilitation of Bergson as a thinker, but also of the goal he had in mind, namely: to develop a critical self-articulation of the hidden philosophical and metaphysical foundations and assumptions of scientific research fields. Here, we should adhere to the adage 'quod supplantandum, prius bene sciendum' or, 'what one wants to uproot, one should first truly know'.

Time is change: a continuous state of becoming – constant renewal. It is not a linear sequencing of infinitely small moments. Einstein restricted his theories to express an interrelation between time and matter, rather than time and change. For him, clock time was the only real time. For Bergson, matter becoming, evolving in concrete and living structures, is the only real time. It should now have become clear that the central notions of duration, vital impulse and immediacy as they were discussed at the beginning of this book, are indeed intimately interrelated. The epistemological value of this interrelation lies in its potential to rethink experienced time as more basic than traditional concepts of time or of space. To fully appreciate this interrelation, we need to take a further step back, namely the step into the reflection of immediate experience. It is at this level that Bergson's ideas meet those of his contemporary and friend, the American philosopher and psychologist William James. James's notion of a 'radical empiricism' takes experience in its immediate sense as the only viable basis. All other knowledge forms are to some extent

© Springer Nature Switzerland AG 2021

L. Landeweerd, *Time, Life & Memory*, Library of Ethics and Applied Philosophy 38, https://doi.org/10.1007/978-3-030-56853-5_8

constructions. Such constructions might be useful for different purposes: they may explain certain events, certain phenomena, but always within a pre-existing embracement of certain knowledge propositions.

Although Bergson displayed a dislike for Kantian philosophy throughout his career, Bergson's ideas can at the same time be regarded as an updated version of Kantian epistemology. What Kant sought to establish was a metaphysical foundation for the empirical validity of Newton's mechanics. To be able to do so, he needed to explain causality. He also needed to explain the nature and status of space and time. It was Kant who posited that space and time are not categories of reality, but rather transcendental forms of human sensibility through which we can arrive at knowledge. Newton had not placed these notions as such. But Kant's transcendental move came at a price: we lost the notion of immediate experience. The noumenon, the thing in itself, became an assumption of the rational mind: it escaped experience.[1] In both the (pre-Kantian) Newtonian paradigm and the Einsteinian paradigm (classical physics and relativity theory), time and space are regarded as real categories, as the categories that determine the basic structure of reality. In contrast with Kant, for Einstein, notions of space and time are realist notions. The distinction between the two is merely one of relationality: where Newtonian mechanics takes time and space as absolute, Einstein sees space and time as intimately related, as spacetime, but this relationship (spacetime) in itself considered as real. Einsteinian relativity informs a worldview that is equally deterministic as Newton's worldview. Kant's epistemological efforts were devoted to safeguard the epistemic validity Newtonian physics, while bracketing its metaphysics. Notwithstanding the radical implications of Einstein's ideas, space and time remained objective categories, rather than being regarded as constituting the scientific way of experiencing and explaining phenomena in reality. Bergson rethought time on a more radical basis. His concept of duration is in a sense no longer related to time, but rather to evolution, change.

The tendency of the human intellect to interpret experience in terms of spatiality render it impossible to understand, intellectually, the phenomenon of change: we can only intuit the experienced flow of duration, and only in ourselves, not in the reality that manifests around us. Intellectual understanding remains dependent of the faculty of language, and due to the conceptual nature of language, change remains beyond the grasp of the intellect. Objectivity in formal scientific discourse is organised around intersubjectively agreed conventions concerning tools, methods and concepts. Legitimacy, validity and objectivity are thus constructs that are by their very nature too restricted to grasp the nature of things.

For the sciences, phenomena like memory, life, and time are often reified in the specific concepts the sciences are able to deal with: brain locationalism (a form of substantivism – see Chap. 5), DNA building blocks (a form of mechanistic reductionism – see Chap. 4) and spacetime (a form of spatialisation – see Chap. 3) are the

---

[1] Although the notion of the sublime in Kant's last critique (*Kritik der Urteilskraft*, or *Critique of Judgment* (1987 [1790]) does allude to a direct experience.

words that make up the vocabulary of such reifications of phenomena as analysable in terms of smaller discrete units. But also within the sciences, many scholars agree that such an analytic approach is not satisfactory: in physics, the complex nature of change is still very much part of contemporary debate, in the life sciences, complexity thought is an important gambit in the discussions over the synthesicing of life, and in the neurosciences complex systems approaches attempt to go beyond the confines of particularism.

To think fluidity is a challenge for the exact sciences. Where physics was traditionally focused on things and their relations (gravitation etc.), it neglected for a long time the phenomenon of change, either ignoring it, or reducing it to a sequence of static conditions. In this regard, biology for a long time focused merely on classification and identification, thus ignoring the processes of change inherent in the evolution of both individuals and species. And Darwinism, in its interpretation of evolution, considered evolution to depend on underlying mechanisms that could be reduced eventually to basic components studied by chemistry and physics. A similar disregard of fluidity can be observed in the neurosciences. In the study of the human neural system, much effort was devoted to the identification of parts of the brain and their function, rather than to the changing and self-creative nature of our neural networks. Mechanical conceptions of the world, be it the world of microorganisms or the world of human behavioural systems and societies, are problematic since they generate rational models which tend to obfuscate systemic change. For societal change, as for any change in complex living systems, a different approach is necessary. Complex coevolutionary systems can only emerge if not controlled from a presumed set of parameters.

How can things possibly be individual when the world is one continuous and fluid whole? A similar question can be raised with regard to time: how can there be a 'present' if time itself is a fluid continuum? Bergson studied this question in relation to the distinctions we experience between past, present and future. These distinctions are quite similar to the distinctions between here, there and yonder, which Martin Heidegger equated with me, you and they. But one cannot equate the now with the here, since to our mind, it can also be 'now' 'yonder', although we ourselves only experience the present in the 'here'. The problem is that these issues are difficult to express in language, because language (the Indo-European languages at least) are biased towards describing the world in terms of entities, individuality and a clear difference between past, present and future. Language is not perfectly poised to express experiences of duration and processes of change.

Usually we consider individuality as singularity. Although things can be 'like' other things', they are unique in their own existence. Things can thus be discerned as discrete from other things. This notion of discreteness is challenged by our experience of space as well as time as a fluid continuum. Thus, continuity and discreteness seem to be opposites: time flows continuously but we immediately experience the present as discrete from past and future. There is a fluid, continuous process which eventually began with the very origin of life, yet, I experience myself as discrete from other beings, even those as close to me as my parents or sister. But, and here things become more complex, am I also discrete and separate from the person

I was when I was 20 years of age? Or the child I was when I was five? Here, in spite of the continuity I experienced in time passing by, I also feel quite different from the blond boy in a blue coat with a woollen hat, holding the feather of a pheasant whilst angrily staring back at me from the pages of my childhood photo album. I may not be the same as that small boy, but I am the same person. I just changed. I grew in length, as well as expanding – in non-negligible terms – in width. When younger I came to be, at times, annoyed by my apparent inertia and my tendency to romanticism – and at times I still do; at other times I got hurt, as we all do; at again other times I surprised myself in how I was inspired to be creative. As such, I may not be the same as that 5 year old, or, for that matter, the 20 year old Laurens that lies in between me and that boy, but I am the same person, since I grew through rather than from these earlier phases. I evolved through these phases to become what I am now, and I will continue to do so until the moment I draw my last breath of air. Whilst most babies act, behave, and are relatively similar, they also already demonstrate different characters, behaviours and tendencies. We are already individual, but we further individuate during our development. This personal evolution is the way how we create ourselves.

In *The World as Will and Representation* (2014), Arthur Schopenhauer discussed the principle of individuation as the basis for multiplicity. Schopenhauer considered space – difference in location – as sufficient ground for individuality and multiplicity. This changed in the psychological use of the term by Carl Gustav Jung. For Jung, individuation describes how we develop an individual self out of a preceding undifferentiated unconscious, as such integrating components of the immature psyche into an ultimately well-functioning whole. Here, it is no longer 'individuality' but 'individuation, seeing the discreteness of things as a result of a process.

Time should be regarded as fluid rather than consisting of discrete elements. This also goes for the interrelation of the sciences. Evolution is fluid, consisting of flows and curves, meandering through the knowledge landscape. But this meandering should not force us into some kind of blurry holism. In *L'individuation psychique et collective*, (1989 [1964]) Gilbert Simondon focused specifically on this process of individuation with regard to technoscience. For Simondon, the individual is an effect of individuation rather than a cause of this phenomenon. It is an ever-incomplete process, and this also goes for the formation of the scientist. Intellectual growth cannot be captured in fixed categories. In this regard, Simondon adheres to the old adage that I discussed in the second chapter of this book: *individuum est ineffabile*. The individual cannot be known. Our rational intellect can merely grasp categories, clusters, taxons, general tendencies etc. while individuality is fluid. Movement and evolution should be regarded in terms of convergence and divergence: discreteness is merely a moment in time, and so is wholeness.

Scientific progress, the articulation of knowledge, from basic insights down to concrete applications, is also a process of individuation. While the exact scientific disciplines that emerged during modernity became differentiated in a process of self-definition and self-differentiation, in our age this process appears to follow a reversed direction, towards convergence into transdisciplinary fields. Moreover, in many fields the current focus of attention seems to shift from basic theory towards innovation in technological systems. Bemoaned by many, applauded by many others, this generated a series of societal transformations of a scale that is without precedent in history, and still requires proper understanding. It is here that the humanities in general and Bergson's insights in particular can play an important role. Scientific progress itself has become an emergent property of complex systems. And as such, the question of how to shape and direct this technological innovation deserves our full attention. After the 1940s, the boundaries between scientific disciplines have eroded. In part, this process is due to the increasing complexity of science: it is no longer possible for individual authorities in the field to combine both the need for a comprehensive overview (generalised expertise) and the need for renewal: science has thus become a team effort, rather than the effort of single geniuses. Science and technology have always been intertwined of course, but their interconnections have become more intimate than ever. Whilst basic understanding through scientific progress facilitated the emergence of technological innovation, innovations of materials, instruments and machines facilitated the emergence of new scientific insights. Anthony van Leeuwenhoek was both a lens maker and one of the first observers of microorganisms. The history of chemistry is riddled with inventors of pigments for paint. Early biologists were researchers of herbal medicine. The nature of this dialectic relation between science and engineering however changed under the influence of mass production of consumable goods.

Scientific research is no longer seen merely as a goal in itself. Societal respect and funding are not based merely on the basis of its potential to answer questions about life, the universe and our place therein. Even astrophysics is expected to contribute to technological renewal. And what used to be merely contingent spin-offs of science and translations of knowledge into technological artefacts is now becoming a central part of the scientific enterprise. In this light, the division of scientific knowledge into different demarcated disciplines is eroding, both via attempts to combine different (formerly irreconcilable) paradigms and via functionalisations that are expected to aid progress in engineering and the manufacture of goods. Scientific thought in various regions of research has increasingly become entangled over the years, notably from the second half of the twentieth century onwards. The convergence of science and technology coincides with a shift of focus from fundamental to applied science. These even applies to fields which were formerly seen as pure, not only astronomy and high-energy physics, but also archaeology, which is increasingly becoming bio-archaeology, conducted with the help of genomics sequencing and other high-tech tools. Astronomy, nuclear physics, archaeology, history, linguistics, all these research areas seem to be converging into theories of everything.

In the sciences, the distinction between 'epistème' and 'technè', between knowing what (knowledge) and 'knowing how' (technological skill) has become blurred. And although in the past, a biologist also needed skills and know how to handle a microscope, it is the technical-epistemic basis of the life sciences itself that is shifting dramatically. The shift in the life sciences from a scientific approach to an engineering approach is, epistemologically speaking, a shift from studying living systems to modifying them. Thus, the scientist himself has become Homo faber – the toiling human, creative and producing – rather than homo sapiens – the knowing human. This carries implies that scientific research has become a high-tech craft. His knowledge is mainly to be found in the dimension of tacit (Polanyi 1958, 1966), ineffable, experiential forms of knowledge, rather than formally articulated theories. An expert in synthetic biology only knows 'that' it works what he or she is doing, not necessarily why it works.

Positivism still is an important aspect of the ideal self-image of science, specifically the view that scientific progress expands human possibilities and thus holds benefits for society. But the assumption that such processes can be predicted and managed into the desired direction is fairly naive. The reverse appears to be true: the emergence of complex technological systems, combined with an economy that drives global masses of consumers into increased consumption and mass production, has resulted in grave global problems. These range from polluted environment up to mental well-being. The emergence of new technologies and the interrelation between knowledge growth, technological systems, and consumer society is a complex development that cannot simply be seen as aiding 'progress' or ameliorating humane human existence. Technological innovation cannot be equated with social progress, while innovation as such is not a guarantee for societal acceptance or acceptability. Whilst some innovations indeed contribute to better qualities of life, higher welfare standards and effective opportunities for environmental care, in most cases the impact on society is fairly ambiguous if not outright detrimental.

We ourselves are increasingly transformed by the technological systems we develop, and our lives, our societal institutions, our communication structures and our ways of positioning ourselves within our worlds are increasingly incorporated into socio-technological systems. Our connection to each other has become a hyper-connection: we develop and display our identities more easily online than in real life, more in interaction with machines and devices (mediated through 'machine-to-machine communication') than in direct, unmediated interaction with our fellow humans and our social and natural environments. Our social relations (as well as our relations to ourselves) are increasingly digitalised (Facebook, e-mail, whatsapp). Also, the boundary between the natural and the digital and between the individual and the collective has become vague, and even our sensory perception of the world has become technologically mediated. We don't experience the world in a direct sense anymore. It is precisely here that Bergson's view become relevant again. In his earlier works, Bergson stressed how consciousness involves immediate awareness of data ('data' as 'givennesses', not in the modern psychological meaning of 'sensory data'). But such an immediate awareness is no longer self-evident. Are consciousness is shaped (or rather: numbed) by technological infrastructures and cultural conditioning. Instead of making us more alert, this numbing of our consciousness seems to have only increased with the advent of new technologies. At the same time, we should not interpret this in a fatalistic manner, for ultimately it depends on *how* we use them. In principle, these technological environments can both enhance and de-enhance cognition. Yet, it is clear that the role of the copy, of the technologically reproducible image, is having an enormous impact on society. Whilst I am typing these sentences, the Amazon rain forest is being burning, but the images of this news event, although continuously recycled via various media, can hardly grasp the magnitude of the event. In this regard, we still fail to realise the impact of our collective behavior on the planet. Yes, we have always been animals without properties dependent on technology, but there is something about these new technologies that is dramatically affecting human nature.

Being human has always meant, being at home in technology. Human nature is inherently technological. But technologies are a φάρμακον (pharmakon), poison and medicine at the same time. And one of the major questions of today is whether we can still succeed in safeguarding an autonomous relationship with the technological systems that surround us and shape our lives: do we facilitate our lives with the help of technological systems, or are our increasingly technologised lives facilitating the emergence and evolution of such systems? In the latter case, we run the risk of creating a world less suited to supporting human flourishing.

We now find ourselves at the crossroads where we can either accept political and ecosystem collapse or opt for transformed creative progress. In our networked, high-tech, information-driven society, we have all become travelling companions. We have to develop a new relationship towards this dematerialised life world. Besides issues of alienation and deterioration, these processes of globalisation also hold unprecedented opportunities. The transitions this world is going through force us to take a different perspective on how we live, act, behave and communicate. This calls for critical reflection and thematisation of our ever-shifting contemporary global societies, established paradigms, and presuppositions, organisational structures, ecosystemic concerns and technological innovations. Here, Bergson's analysis of technology as an intrinsic and fluid part of human nature should appeal to our current ways of thinking about technology, since too often, these ways of thinking remain mechanic in nature.

8.1 Thinking Fluidity

Bergson's work confronts us with an important question: is it possible to 'think fluidity', liberating ourselves from the vocabularies and metaphors that constantly refer to what is fixed and solid? Can our words and ways of thinking become more susceptible to the fluid, dynamical nature of the world as a creative evolving process, or do we only have access to a restricted vocabulary, one that steers our thinking into understanding the world in terms of static conditions? The problems we are facing are also problems of the intellect, the nature of human thinking: our ways of perceiving, experiencing and imagining our world.

It is quite apparent that Bergson aimed to criticise this core tendency of western philosophy. The problem he identified concerned the dominant tendency to interpret the world as 'a collection of things' (Bor 1990). This tendency resulted in (and became quite apparent in) the dualistic metaphysics of René Descartes, but it continued to influence western philosophy ever since. This tendency is not merely be a product of a history of ideas, it is a product of language as such. Language as such may be deficient, and the deficit might even be irreparable.

The problem of language is not restricted to occidental philosophy. It is also addressed in oriental philosophy. In Chinese thought, we encounter it in the duality between Confucianism and Taoism. In relation to the political structure of society, Confucius held that language (names) should correspond "with the truth of things" (Ware 1980). In Confucius' worldview, language suggests the existence of order. Thus, if we fail to call things by their proper names, we will fail to come to grips with reality. Names should be rectified and clarified, to correspond with the nature of reality. Here, Confucianism seems quite congenial with the classic Greek notion of logos, which means word, discourse, reason, relation, ratio and cosmic law. Taoism, however, holds that the flow of things cannot be captured by naming it, since "the names that can be given are not adequate names", as we are told in the opening sentences of the Tao Te Ching (Legge et al. 1891a, b). Reality cannot be grasped by language, and Zhuangzi wonders whether language is even different from the chirping of birds. So, also in the Chinese worldview a schism seems to exist between the belief in the possibility of a parallelism of names and things and scepticism concerning the ability of words to capture the dynamic nature of things. What exactly is the difference between the chirping of birds and the words in a book? Although words may be ultimately ill-equipped to grasp the nature of the world, they may nonetheless be able to allude to it. This is one of the reasons why Bergson's philosophy, in spite of its literary elegance, is often regarded as obscure. He wanted the words he used to adhere to the creative flow of his ideas. He could not express his views via concepts and definitions. Instead he wanted the words to speak for themselves, because explaining them with the help of a static vocabulary would obliterate their initial function. Bergson was, paradoxically perhaps, an author with a chronic distrust in language. This effort to create a more fluid vocabulary is not only beneficial for philosophy: it will also benefit the sciences, because the current scientific language is still infected if you like by an outdated metaphysics. To make science more sensitive to the creative fluidity of nature requires not only technological innovation, but also a drastic transformation and critical reconsideration of the languages of science. This would benefit society as well. The implication of Bergson's view on language is that we have failed to address processes such as climate change in a timely manner because of language as our main obstacle. The development, not only of new forms of research but also of more effective global policies, requires a linguistic transition from static terms and metaphors towards more fluid forms of language. Sometimes, this might entail leaving behind the notion of language as expression of potentially valid knowledge, trading it instead for more poetic and artistic ways of expression.

The focus on fluidity also explains why, in his own time, Bergson was so often misinterpreted. As mentioned earlier, Bertrand Russell strongly opposed his views, but without grasping their significance (Vrahimis 2011; Russell 1992, p. 319). Notably, Russell did not grasp Bergson's assertion that notions of continuity in philosophy, mathematics and physics are distorted by our cognitive and linguistic habit to translate these into spatial terms, hindering us to see them as creative processes.

Instances are reduced to points on a line whilst the temporal nature of continuity is neglected (Petrov 2013). Bergson's struggle with the problem of language might have been partially responsible for these misunderstandings. Although Bergson himself did not yet use terms such as individuation or autopoesis, this vocabulary may allow us to better understand his ideas. Ultimately, the current environmental crisis results from the fact that reality itself does not adhere to our ingrained cognitive categories. Here, only intuition can offer alternative ways of exploring and experiencing the world.

8.2 On Dualism

One of the benefits of fluid thinking as outlined above is that it allows us to move away from dualistic worldviews of the past. The doctrine of dualism builds on the idea that two substances determine the fabric of reality: material substance (extended into three-dimensional space) and thinking (as something which somehow exists outside of this material, spatial realm). Humankind is supposedly involved in both (as a corporeal thinking being), whilst the world of things is only made up of matter. At the other end of the spectrum, the divine is a purely thinking substance. Humans are, according to this doctrine, both material and spiritual. We are bodily entities consisting of skin, veins, sinews, muscular structures and a lot of gooey stuff inside. But we can also remember, imagine, reflect, conclude, dream etc. According to dualism, we are thus both body and mind, matter and spirit, earthly and divine. This doctrine is riddled with contradictions, as was discussed in Chap. 5. Why, then, is this doctrine so tantalising?

We are part of the world that appears to us, but at the same time we have a perspective on that world, as if from some privileged outside position. Our existence is neither fully integrated in the processes of the universe, nor fully independent and outside of those processes. This dualism might best be illustrated through the words of a contemporary of Bergson, the Indian poet and philosopher Rabindranath Tagore,[2] who expressed this dualism (which plays a role in both eastern and western thought) very elegantly in book II of *Sadhana*, in the chapter on the Self:

> "At one pole of my being I am one with stocks and stones. There I have to acknowledge the rule of universal law. That is where the foundation of my existence lies, deep down below. Its strength lies in its being held firm in the clasp of the comprehensive world, and in the fullness of its community with all things. But at the other pole of my being I am separate from all. There I have broken through the cordon of equality and stand alone as an individual. I am absolutely unique, I am I, I am incomparable. The whole weight of the universe cannot crush out this individuality of mine. I maintain it in spite of the tremendous gravitation of all things. It is small in appearance but great in reality. For it holds its own against the forces that would rob it of its distinction and make it one with the dust."(Tagore 1915, p. 40).

---

[2] Bergson was the first president of the International Committee on Intellectual Cooperation (ICIC), the predecessor of UNESCO. Rabindranath Tagore was a liaison of ICIC.

So, although we might artificially generate a cosmological bird's eye perspective, with the help of technology for instance, we cannot but acknowledge that it is only from such a perspective that we seem insignificant, part of stock and stones, whilst on the other hand, the very reason why we can actually take a position of a self, sets us apart from stock and stones. This uniqueness is, as Tagore continues:

> "[…] the superstructure of the self which rises from the indeterminate depth and darkness of its foundation into the open, proud of its isolation, proud of having given shape to a single individual idea of the architect's which has no duplicate in the whole universe. If this individuality be demolished then though no material be lost, not an atom destroyed the creative joy which was crystallised therein is gone. We are absolutely bankrupt if we are deprived of this speciality, this individuality, which is the only thing we can call our own; and which, if lost, is also a loss to the whole world. It is most valuable because it is not universal. And therefore only through it can we gain the universe more truly than if we were lying within its breast unconscious of our distinctiveness. The universal is ever seeking its consummation in the unique. And the desire we have to keep our uniqueness intact is really the desire of the universe acting in us. It is our joy of the infinite in us that gives us our joy in ourselves." (Tagore 1915, p. 40).

So, essentially, we experience ourselves as an 'I' and at the same time know that we are part of the world that this 'I' perceives. We are thus both a thinking being and a material being. Hence, we seem caught up in dualism, in spite of its many pitfalls. Can we overcome this problem? This is possible only if we step back and return to our initial intuition of the world as duration, as an evolving process. Our given assignment then becomes to try and develop a language more suitable to expressing this experience. But expression in language poses a problem. Only the genres of the imagination can be of help here: poetry, theatre, the visual arts etc. are less restricted by discursive conventions, and convey something beyond intellect. In a sense, philosophy might better be placed at academies of art than at universities. This would not only allow for the development of a more viable philosophy, more equipped to overcome classical dualism. It might also further the evolution of less destructive and disruptive ways of life.

8.3   Upgrading the Human

Bergson also changes our perspectives on other societal and technoscientific discussions, such as the one on human enhancement: the idea that humankind needs to step beyond the confines of natural evolution. Research fields such as genetics, genomics, pharmaceutics, prosthetics, etc. carry the promise to to realise this step beyond nature. Given the deficits of human nature, enhancement might even prove inevitable for our survival. Some transhumanists appeal to the notion of extropy (More 2003). The idea is that in our cognitive and technological ability to consciously steer our own evolution will allow us to redirect our evolutionary self-adjustment away from the supposedly entropic aspects of 'natural' evolutionary processes. Similar to negentropy, a concept we have discussed earlier (Schrödinger 1944), extropy bears a resemblance Bergson's concept of vital impulse (élan vital), but whereas Bergson merely aims to *understand* the flux of emerging technologies, transhumanism jumps towards *prescription*: we have a 'duty' to self-enhance. In this transition, however, transhumanism makes a crucial mistake, namely by adopting a mechanistic approach, by seeing humans as material substances extended in space, waiting to be technically improved, rather than as evolving, complex, living, self-conscious organisms. While transhumanism conceives of humans as technically controllable, Bergson sees humanity as a complex fluid being evolving in time. The fatal flaw of transhumanism, its dualistic and mechanistic view on human existence, is engrained in the language, in the vocabulary which the proponents of transhumanism have adopted. As a result, transhumanists forget the unpredictability of processes of emergence of technological systems.

Genealogically speaking, transhumanism is related to Social Darwinism, an intellectual movement which can be summarised as a two-sided thesis (Kitcher 2012): first there is the idea that humans (as individuals) have intrinsic abilities and talents, expressed in how they act and what they achieve and second, there is the idea that we need to intensify competition so that the most talented can develop their potential to the full. Ultimately, everybody benefits and weakness is cancelled out. In other words, Social Darwinism is an individualist 'survivalist' view (such as the one developed by Rand (1992[1957])), emphasising survival and individual flourishing on the basis of egotism, which can also be discerned in transhumanism. Social Darwinism continues to inspire deliberations concerning ethics, human self-understanding and biology (Spier 2006) but again, its view on nature, evolution and human beings is basically flawed, perceiving biological systems as controllable, makeable and predictable. Human enhancement is more than merely the aim to use medical technologies. These are usually developed for curing and preventing diseases, not for enhancing the human being, but the line between these two goals is often blurry (Landeweerd 2009). Those adhering to the idea of a biological self-adjustment of the human species often remain caught up in the mechanistic view on human existing.

The positivist expectations prevalent in the human enhancement debate are overly optimistic about the effectiveness of human responsibility with regard to technology, and remain naive by seeing the nature of technological innovation as predetermined. Expectations concerning innovation in robotics and artificial intelligence lean heavily on the presumption of a material and technological determinism. These discoveries that are now applied to human cognitive functioning seem to strengthen the view of the human mind as a clockwork that can be recalibrated at will. Talent, IQ, concentration, empathy, they all seem to be construable on the basis of insights into the biochemical fabrics of the brain and a whole field of experts have proclaimed the emergence of a post-evolutionary era. Cognitive enhancement is firmly established as a research field in the boundary zone between the natural sciences and the humanities. But this builds on a fatal misconception of human cognition as such. From Bergson's perspective, proponents of human enhancement technologies are traditionalists. They miss the point that humanity has always already transcendeds the DNA-restricted evolutionary processes that govern the rest of nature since time immemorial, because the human condition temporal fluidity is inherently technical, (althoughbeit not mechanical). We constantly create new ideas and needs and as such we are beings that transcend our genetic nature continuously. Ideas and concepts should be regarded as life forms. This includes our notions of self. We constantly need to create new concepts, methods and strategies to reconstruct ourselves (Landeweerd 2009).

8.4   Internationalism and the Exchange of Worldviews

As I discussed in Chap. 7, in 1922 Henri Bergson became the first president of the *Commission Internationale de Coopération Intellectuelle* (CICI – the International Committee on Intellectual Cooperation).[3] It was founded by the League of Nations (predecessor of the United Nations), on the basis of a French proposal to establish a platform for international intellectual collaboration. The CICI (also IICI) was to perform as the League of Nations' advisory committee. Paralleled by *the Institut International de Coopeération Intellectuelle* (established in 1926), the CICI's aim was to promote international exchange between scientists, researchers, teachers, artists and intellectuals. Famous individuals associated with the CICI included commission members such as physicists Albert Einstein, Marie Curie and Hendrik Lorentz, politician and statesman Sarvepalli Radhakrishnan, biologist Kristine Bonnevie and consultants such as poet Paul Valéry, composer Bela Bartok and author Thomas Mann.

The League of Nations Council took the position that CICI members were appointed in consideration of their personal ability and their reputation amongst peers, but without any discrimination as to nationality (Greaves 1931; Laqua 2011). The CICI had a different scope and purpose compared to its later successor organisation. UNESCO seeks international collaboration on education, the sciences and culture. It functions as a diplomatic entity. The CICI was more focused on specific individuals who were praised for their academic scholarly or artistic activity but at the same time committed to public intervention and international collaboration (Laqua 2011).

The interbellum was a period of unprecedented enthusiasm over cultural internationalism in the sciences, the arts, literature and diplomatic collaboration. These elite circles were apparently unaware of the dramatic long-term effects which the conditions imposed on Germany through the Treaty of Versailles were having and there was much optimism concerning the idea that nineteenth century nationalism and imperialism, important causal factors triggering World War I, would gradually be replaced by a peaceful international community. This necessitated institutions that would support the creation of such a new, post-nationalist order, and this called for an intellectual countermovement aiding the identification of global moral, intellectual and spiritual values. Still, complete post-nationality was not achieved in the

---

[3] Sometimes also referred to as the League of Nations Committee on Intellectual Cooperation

CICI either. The position of Germany remained controversial. A motion by the right winged nationalist Swiss CICI-member Gonzague de Reynold to normalise the relation with German intellectuals was strongly opposed by Bergson, who believed it was too early to establish normal relations with German scientists so soon after the end of the war (Laqua 2011).

In its focus on education as a means for global pacification, the CICI also sought for exchanges between different traditions in the world. Gilbert Murray for instance collaborated with the aforementioned Indian poet and philosopher Rabinadrath Tagore. Tagore had established Visva-Bharati, an institute that explored the exchange between eastern and western worldviews. Together with Murray, Tagore authored a volume in the IIIC's series 'civilisation'. Thus, CICI and IIIC sought a civilisational dialogue through 'ambassadors of culture in different countries or groups of countries with different cultures'. International education was to go hand in hand with cultural exchange. Only then, global peace might be achieved. Bergson's focus on international peace was eventually best articulated in 'The Two Sources of Morality and Religion' (1932).

Now that we are entering a new era, the role of international organisations and platforms is shifting as well. Against this backdrop, a rereading of Bergson's book may inspire a reconsideration of the role of UNESCO and similar organisations. Under the pressure of the challenges we are facing, can a global bioethical culture now really be achieved as the outcome of a process of creative moral evolution from group morality to universalism? Whereas the hope initially was that new communicative structures (Internet, WWW, etc.) would facilitate convergence and the emergence of global podiums for deliberation and convergence, new media now rather seem to incite to divisions and collisions: a relapse into group-think. Is Bergson's position outdated or more relevant than ever? As indicated, the attitude of hope versus cynicism or despair suggests the latter alternative. This time needs a rival discourse, more adequate when it comes to addressing global forms of crisis than the neo-liberal vocabularies of innovation, seeing growth in spatial terms: in terms of expansion, rather than in temporal terms, in terms of a progress of the human soul.

8.5 Innovation and the Market

Innovation is a term that is often used in discussions about alignment of scientific research and the global market. In this debate, however, a rather specific view (or rather: language) is dominant, namely neo-liberalism. And liberalist views on the market a quite congenial with Social Darwinist assumptions. The market is assumed to function as a natural ecosystem in which the evolutionary principle of survival of the fittest applies: ecosystemic economic conditions and product design would then be the factors that determine to what extent a producer survives on the market. What survives is good, and what is good survives. The tacit assumption of this approach to the market is the idea that the market automatically follows the path of progress. This generates a vocabulary in which interventions ('disturbances') will bar human social progress, whilst not intervening in the mechanism of the market will eventually be beneficial for all. Bergson already commented on this: "Increasing experience has proved [...] that the technological development of a society does not automatically result in the moral perfection of the men living in it, and that an increase in the material means at the disposal of humanity may even present dangers unless it is accompanied by a corresponding spiritual effort" (Bergson 1928). What seems to be lacking in the dominant, neo-liberal views on innovation is a critique of its direction towards mere expansion. A critique of the short-comings of the innovation paradigm requires a re-examination of the interrelation between time, nature, technicity and human evolution. Science today is different compared to science in the early twentieth century. It has evolved from small-scale curiosity-driven research to application-oriented innovation. Where Niels Bohr might be typified as a curiosity driven researcher par excellence, Alexander Graham Bell may be seen as the scientific engineer, but in contemporary research, scientists such a Craig Venter represent a convergence of these extremes. This marriage between fundamental research and technology innovation coincides with a more fundamental marriage of natural evolution on the one hand and the emergence of powerful technological systems on the other. A lucid assessment of the meaning of these developments requires perhaps not the marriage of but at least a renewal of the dialogue between philosophy and science, a dialogue which was unfortunately disrupted due to Einstein's earlier and fatal misunderstanding of Bergson's views. Such a dialogue holds the promise of intellectual renewal for both fields. Both philosophy and science will be challenged to reconsidered their basic convictions and vocabularies.

8.6   Brief Postscriptum

As indicated in this study, the starting point for this renewal basically consists of three views: the identification of time with change; an extension of the notion of evolution and the positioning of freedom as positioned between memory and imagination rather than between determinism and indeterminism. Traditional metaphysical and scientific understandings of substance and time should be replaced by they insight that all sciences ultimately address the same phenomenon, namely change. Natural, technical and political evolution should be reconsidered from this perspective. This means that we should see biological, DNA-based evolution (a process of variation and recombination) not as different from, but as intimately entwined with similar processes of creative growth in the technological and noological realm. Growth is not expansive, but creative. The gradual "emancipation" of the living from the non-living and of the spiritual from the material is by no means a miracle. Human self-consciousness and freedom as we experience them should not be seen as static attributes (as 'things' which we allegedly 'have' or 'possess' in contrast to other species), but rather as the temporary outcome of processes of remembrance and imagination, evolving in fluid time.

Literature

Bergson, H. (1928). *Banquet speech for the Nobel banquet* (given on Bergson's behalf by ambassador Armand Bernard). https://www.nobelprize.org/prizes/literature/1927/bergson/speech/. Last accessed, 12-10-2020.

Bergson, H. (1999 [1922]). *Duration and simultaneity* (eds: Durie, R.), Manchester: Clinamen Press.

Bergson, H. (1977 [1932]). *Two sources of morality and religion* (trans. Audra, R. A., Brereton, C.). Notre Dame: University of Notre Dame Press.

Bor, J. (1990). *Bergson en de onmiddellijke ervaring*. Meppel: Boom.

Greaves, H. R. G. (1931). *The League committees and world order: a study of the Permanent Expert Committees of the League of Nations as an instrument of international government*. London: Oxford University Press.

Kant, I. (1987). *Critique of judgement* (trans. Pluhar, W.S.). Indianapolis: Hackett Publishing Co.

Kitcher, P. (2012). The taint of social darwinism. *NY Times Article*. http://opinionator.blogs. nytimes.com/2012/04/08/the-taint-of-social-darwinism/

Landeweerd, L. (2009). *Reconstructing the self: Issues of choice, fate and justification in the eugenics debate*. Maastricht: Datawyse.

Laqua, D. (2011). Transnational intellectual cooperation, the League of Nations, and the problem of order. *Journal of Global History, 6*(2), 223–247.

Legge, J., et al. (Eds.). (1891a). *The Tao Teh King, Sacred Books of the East, Vol. XXXIX*. Oxford: Oxford University Press.

Legge, J., et al. (Eds.). (1891b). *Sacred books of China, Vol. V*. Oxford: Oxford University Press.

More. (2003). *Principles of Extropy (Version 3.11): An evolving framework of values and standards for continuously improving the human condition*. Extropy Institute. Archived from the original on 15 Oct 2013.

Petrov, V. (2013). Betrand Russell's criticism of Bergson's views about continuity and discreteness. *Filozofia, 68*, 10.

Polanyi, M. (1958). *Personal knowledge: Towards a post-critical philosophy*. Chicago: University of Chicago Press.

Polanyi, M. (1966). *The tacit dimension*. Chicago: University of Chicago Press.

Rand, A. (1992 [1957]). *Atlas shrugged* (35th anniversary ed.). New York: Dutton.

Russell, B. (1992). *The collected works of Bertrand Russell: Logical and PhilosophicalPapers, 1909–1913, vol. 6* (p. 319). New York: Routledge.

Schrödinger, E. (1944). *What is life – The physical aspect of the living cell*. Cambridge: Cambridge University Press.

Schopenhauer, A. (2014). *The World as Will and Representation*, Volume 1&2 (trans. Norman, J., Welchman, A., Janaway, C.). Cambridge: Cambridge University Press.

Simondon, G. (1989 [1964]). *L'Individuation psychique et collective*. Paris: Aubier.

Spier, R. (2006). Evolution and ethics: Is an evolutionary ethics possible? In L. Landeweerd, L. M. Houdebine, & R. ter Meulen (Eds.), *BioT-ethics an introduction* (2nd ed.). Florence: Pontecorboli.

Tagore, R. (1915). *Sadhana, book II: The problem of self*. New York: MacMillian.

Vrahimis, A. (2011). Russell's critique of Bergson and the divide between analytic and continental philosophy. *Balkan Journal of Philosophy, 3*(1).

Ware, J. R. (1980). *The Analects of Confucius*, Book 13, Verse 3. Pennsylvania: Franklin Center.

Acknowledgements

Hypocrite lecteur!—mon semblable,— mon frère
   - Baudelaire, preface to 'Les fleurs du mal'; quoted in T.S Eliot's 'The Wasteland'

This study is the result of a relatively recent choice to synthesise the pragmatic work on the place of science in society that characterises my current academic role with the nature of my original studies in philosophy: whilst I never abandoned this basis, it became increasingly difficult to retain my original ambitions in the field of metaphysics and epistemology in the pragmatic contexts of policy advice on science, science and society interactions, ethics of science etc.. At the same time, had the occasion arisen to opt for a career at a faculty of philosophy, I would not have felt comfortable with the idea: in philosophy, many churches and belief systems compete, and often, no true impact is made on the world: would I have opted not to enter into the studies that resulted in this publication, and would I have spent the past 20 years dabbling in metaphysical theories, I would have missed many inspiring contexts. I would also have neglected an obligation to peer communities in the ethics of science, the philosophy of science, the sociology of technology by leaving the context of science. What I outlined in this book may not always be evenly recognised as important by colleagues working in these fields, but the lack I felt in mine and their practices is one of the driving forces for entering in this endeavour. Had I not entered it, I would also have been guilty of disrespect to experts from those domains in science that welcomed my collaboration over the past two decades.

I believe that the result of my studies is helpful to either the academic domains that investigate the praxis of the exact sciences, or the domains of those exact sciences themselves. I also believe I managed to take some steps in philosophy: this study followed from discussions, meetings, and collaborations of a very broad and interdisciplinary nature: I am indebted to scientists, engineers, policy makers and also philosophers for providing me with an extremely rich palette of viewpoints, insights and advice. This resulted in what I termed an 'applied metaphysics': a metaphysics for a living and current world that faces problems of such a magnitude that existentialism has become a geological category. But this book will not change worlds: steps to a translation from this 'applied metaphysics' to pragmatic contexts

L. Landeweerd, *Time, Life & Memory*, Library of Ethics and Applied Philosophy 38, https://doi.org/10.1007/978-3-030-56853-5

are not realised by the mere writing of a book. Its intentions, if they are read, call for action. This means that I need to express my acknowledgements first and foremost to you: the, in the words of Baudelaire, hypocrite reader who is my double and my brother. Books lend themselves to many forms of reading: one might hop from one paragraph to another, go through these ponderings, assessments, analyses etc. from footnote to chapter, and from reference to quotation without following the supposed red line that makes up their composition. I did not read my authors very differently. Being a hypocrite reader myself, I am indebted to you.

I would like to thank Hub Zwart for his intuition that my experimental approach to a reading of Bergson in the context of current and past scientific theories might yield more than what initially would have been a mere introductory book chapter: it is a bit of a leap of faith to have someone write a book that is based on an expertise that still needed to be established with regard to the main philosopher in question. Further, I am indebted to Hein van Dongen for our open dialogues and critical advice. Pieter Lemmens lent his wisdom to my views in his expertise on the role of technology in the anthropocene. My conversations with Pieter L. Janssen were of elementary importance: our in depth discussions on the subject matter of this book as well as on the problematic choice to write this book in a language that is not my first truly helped sharpening my mind and words.

Julie Boonekamp's critical remarks on my initially sloppy treatment of Husserl's notion of Epoche were more than helpful, as was her patience with my monomaniac self-referentiality in our conversations over these past three years. The friendly circle of experiential experts in phenomenology and psychiatry that she led in Amsterdam drew me from my hermetic lifestyle: I would also like to thank Sjoerd van Tuinen to take up the position of co-promotor at such a late stage. I am indebted to, Henk de Regt for his comments and deep reading of a semi-final version, Simon Butter for his laborious work on my English – a second language to me – and his work on my messy references. I also want to express my gratitude to my colleagues of Filosofie Oost-West for their patience in their reading of an at that time less readable draft of Chap. 5, my colleagues at iArts for their enthousiasm, my colleagues at the department of ISiS, Radboud University Nijmegen, for bearing with me in this at times somewhat solistic enterprise. I would also like to thank Jan Bor, for abstaining from interfering with this four year process – merely lending one of the two copies he still had of his study of Bergson's oeuvre. I remain unsure whether I will return it.

Ray Spier is someone who deserves special mentioning. Although we sometimes clashed on personal as well as professional levels, he was a dear colleague and a friend. He lent his eyes to a reading a first draft of the master text. He spent his time on what still amounted to a messy rubble of fragments and late night associations. In spite of its then state, I am glad to have sent it when he was still alive, and regret the fact that he is no longer there to read the final result. Our at times averse views on the nature and desired role of biotechnology were an important source of inspiration in earlier stages of my career, and I remain grateful for having known him.

Lastly, in the years that I wrote this book, I lived in an apartment with a view on hills and cows in meadows, on a border between two countries. I wrote with the noise of a crossing of seven roads on which an old linden tree stands. I would like to thank that tree, those roads, that view, that noise, those cows peacfully chewing their grass in those meadows.

Afterword

The reason for writing this book is not a purely academic one. It also attempts to unveil a somewhat neglected worldview. Some fragments of this worldview reached me via my close encounters with friends now deceased, of the generation of my grandparents. I miss them. I miss the dialogue I was able to have with them, one that spoke of politics, art and family in an age that is no longer alive, except for a chandelier given by one, or an anecdote over an encounter with a famous artist by another. I miss the generation that came of age before the second world war started. I miss the strange amalgam between mourning and experiment; that post-aristocratic nineteenth century mentality in which the traditional uniform of an undertaker became the general dress of gentlemen, as if the whole of Europe immersed itself in mourning over the decapitation of a French king. I believe this worldview should not disappear with its historical witnesses: let us rather unseal the tombs (jealously guarded by scribes) in which this vibrant worldview came to be enshrined.

Early on in my career, or, to put it more properly, before it even started, I experienced a 'eureka' moment which at the same time was an aporia – an impasse: I saw the spectrum of philosophy as determined by two extremes that should be avoided: on the one extreme, there was the solipsistic conclusion: that the world appeared to be nothing more than a figment of the imagination (and, by necessity, my own imagination). That which I had come to know as 'the world' might, in that case, have been planted there by a God-like entity of dubious morality. At best, the world would be something that was – to cite Arthur Schopenhauer – merely my representation, a conclusion that appeared to be repugnant in many ways. On the other extreme there was the materialistic conclusion: that even my own sense of self was nothing more than a bundle of causally determined neurochemical patterns, in no way different from the things that appeared in my sensory perception. My 'eureka', which was in all likelihood not unique, consisted of the idea that it should be both. Whilst seemingly a paradox, the world was me, and I was the world.

© Springer Nature Switzerland AG 2021

L. Landeweerd, *Time, Life & Memory*, Library of Ethics and Applied Philosophy 38, https://doi.org/10.1007/978-3-030-56853-5

When alone on a mountain ridge in the French Pyrenees – unaware of the parallel to Nietzsche's similar experience on a hill near the Italian coastal resort Rapallo[1] – I chose not to become a goat shepherd there, nor to delve into a career as an artist. I continued my studies in philosophy instead. But I did remain involved in the visual arts, photography, cinematography and theatre. I sensed something in the postmodern wave in philosophy, which was still quite dominant at the time, that was at odds with a basic intuition: its linguistic idealism. I found friendship with philosophers and scientists of a different mind and an older generation that predated or were unrelated to postmodernism – with them, I could better share my views than with the generation of my professors; I was cautious with exploring western views of oriental philosophy, due to their often Platonic and dualistic explanatory tendencies (matter and spirit, things and souls) and I tried to refrain from theology due to its conventionalist dogmatism for similar reasons.

In the end, I wound up working predominantly in European advisory and participatory projects on new and emerging science and technology, including human genomics, neuroscience, industrial biotechnology, synthetic biology and nanotechnology – with a specific focus on ethics, regulatory frameworks and socioeconomic drivers of innovation. But although I found snaps and bits in art, poetry and literature, my philosophical intuitions remained in a state of slumber. It took some twenty odd years to find the right vocabulary to express my intuited golden mean position in philosophy.

It took until my career brought me to Radboud University Nijmegen for paths to converge again. Both my discussions with scientists and my discussions with colleagues devoted to philosophical reflection on science (which still holds a relatively protected position there) enabled me to find my language –albeit predominantly in English, not my mother's tongue. Still, I avoided the study of the works of Henri Bergson since the study of his works seemed fairly time consuming, and time was not a commodity I had abundantly available. But when invited to contribute to a book that aimed to revitalise a number of neglected French philosophers of science by the head of my department, I could no longer avoid the curiosity that had been aroused by the discussions I had had about Bergson – specifically since no other sufficiently neglected Frenchmen that held my interest came to mind.

As a philosopher, I mainly worked on projects funded by the European Commission on science in society. This work involved the ethical issues carried along by scientific research, as well as its regulaton, its policy and ultimately its goals, yields and deficits. This work focused on topics such as genomics and health, genetics and agriculture, nanotechnology and industry, biotechnology and international trade etc. I did not find a common ground in terms of the type of philosophy that applied to these various fields. My work mainly involved critical discourse on agency and prenatal diagnosis, naturalness and genetic modification, distribution of wealth and international trade, or neuroimaging and human consciousness. But

---

[1] It was there that he conceived of Zarathustra. I was also unaware that I would give my first public lecture in a conference venue on that same hill in Rapallo some years later.

when invited to elaborate on parts of the findings of my past career in terms of a chapter on a French philosopher, I intuitively decided to explore the works of Bergson, whose works I had more or less avoided before, due to the pressure of the mandate that some friends and colleagues that were more initiated in his oeuvre professed, and seen the reluctance I had to devote my time to one thinker[2]. After all, why become an adept of a grand name of the past, if one should be an autonomous thinker, even if thinking then results in tinkering.

In spite of these ponderings and old and possibly false self-commitments, I became increasingly overwhelmed by the richness and subtlety of Bergson's thought. The relevance of his work for my initial intuitions, and this study also helped to overcome some persistent problems with postmodern philosophy's treatment of language and symbols[3]. I stumbled upon several authoritative publications that already covered the relevance of Bergson for science. But I found none that focus on the fields that I had been involved in during my work for the abovementioned European projects on science in society. And in spite of the bureaucratic nature of such projects, these did from me as a researcher: the attempt to create dialogue across scientific cultures, domains and traditions, the need for collaboration in defining the boundary between vision and propaganda, the value of reflection in arenas more designed for mere argumentation. It is for this reason that I wrote this book.

This publication aimed to extrapolate Bergson's core ideas to newly emerging knowledge areas and their application in various fields of technological and societal innovation. My work on Bergson is still 'work in progress', as it should be if we follow the process-philosophical notion of constant renewal. But a critical saturation point to publish this book for the audiences that may find use of it is reached since a more in depth account of my analysis would become overly specialist.

Whilst this book does shed some new light on specific aspects of Bergson's oeuvre, it does not aim to provide for a full historic interpretation of his oeuvre. As such, I did not aim to provide for an addition to the existing library of Bergson monographs, since my ambition is not aimed at the field of the history of philosophy. I rather seek to realise a recollection and revitalisation of the ideas and views of Bergson in their potential relevance for scientists and philosophers today.

---

[2] And also after the writing of this book, I am not inclined to call myself a 'Bergsonian thinker'. The study of his works and those commenting it merely gave voice to the multiplicity of what I had already been doing before. Still, I do have to acknowledge that I might not have found such voice without reading Bergson, without reading on Bergson.

[3] After Thomas Kuhn's definition of knowledge paradigms, the idea of universal validity came to be problematic, parring sociological accounts of knowledge systems as a 'mere' sociocultural phenomenon against exact scientific adherence to the possibility of valid knowledge of phenomena in the world. The sociocultural reductionism took place in parallel to the emergence of postmodern relativistic views on knowledge growth. The 'anything goes'-argument following from this postmodern worldview remains unsatisfactory and, in spite of any true renewal of a 'realist' worldview, we live in an age 'after' postmodernism (van Peursen 1994).

Conventions Pertaining References, Language and Terminology

Conventions used by any book should aim at consistency rather than purity. There is no way to determine correctness in referencing. Some conventions however do deserve attention. There are cultural, linguistic and grammatical reasons to divert from the standards of a specific referencing style. This book makes use of the Chicago referencing style, but it diverts from this style in several specific cases. These concern usage of concepts from other languages than English, the use of other alphabets (Chinese, Greek) or other variants of our alphabet, and the usage of other languages. For this publication, the general line is Oxford styled English, albeit without the luxuries of a native speaker in terms of the imaginaries that are at hand for a native speaker. As such, some phrasings in this book are impoverished whilst other ways of thinking that could only emerge from the use of English from the perspective of another language may have been gained.

Key writings originally published in another language are referred to in English, whilst referring to the original titles in their first use, and at the beginning of each new chapter. Each time the original title is used with no special reference, the original date of publication is mentioned as well as the publication in English that was used. Wherever specific quotations or reference to specific pages are used, the year of publication of that specific translation is also used, again, next to the first year of publication of the original first print, or, exceptionally and only when needed, reprints that contained relevant amendments or additions.

Key writings with a historical significance are referred to by the publication and year that were used, whilst referring to original dates of publication between brackets. For key works predating modern standards (e.g. Greek or Chinese antiquity), such conventions were not used.

In reference to Greek philosophical terms and phrases, English terms were used followed by their Greek original use between brackets. One exception is the usage of the Aristotelian four causes, which are traditionally styled in Latin due to the Aristotelian influence on medieval philosophy (which was predominantly in Latin

L. Landeweerd, *Time, Life & Memory*, Library of Ethics and Applied Philosophy 38, https://doi.org/10.1007/978-3-030-56853-5

for western Europe[1]. In general, the rendition of latinised versions of Greek words is designated with hyphenations and other punctuation marks, to an extent relicts of presumed pronunciation.

For proper nouns: Dutch, French and German last names at times contain words that are treated as part of the name proper in English. Due to the confusions that might arise for native readers from these countries, this book opts for a retaining of Dutch convention for Dutch last names (containing 'van', 'van de' or 'van den'), French convention for French last names ('de'), and German conventions for German last names ('von'). These interfixes designates places of origin, whilst the last names in themselves are the second, and in some instances third word in the last names. As a result, the literature list lists for example 'van Dongen' under the 'D', as 'Dongen, van', or 'van den Belt' under 'Belt, van den'. Further, capitalisation of these 'infixes' is not the convention for either of these languages. Therefore, for example: 'Teilhard de Chardin' rather than 'Teilhard De Chardin' (and in the litera-ture list, following French convention, not 'Chardin, de, Teilhard, P.', rather 'Teilhard de Chardin, P.').

---

[1] Although from the early seventh century the Byzantine empire increasingly turned to Greek as its Lingua Franca under the reign of Heraclius (r. 610–641), when the Empire's military and adminis-tration and adopted Greek for official use instead of Latin.

Glossary

Abiogenics: The study of the evolution of early life on earth, focused on how organic molecules and subsequent simple life forms first originated from inorganic substances.

Autopoiesis: Creation out of itself, spontaneous emergence. Introduced in 1972 by the Chilean biologists Humberto Maturana and Francisco Varela to define the self-maintaining chemistry of living cells. The concept is also used in the cognitive sciences, systems theory and sociology to describe how certain systems are able to maintain themselves and reproduce.

Brain locationalism: The theory that posits that concepts in the mind can be found in specific parts of the brain.

Clock time: See: Temps.

Complexity: As opposed to complicatedness, complexity is a concept that accounts for change and flux. See: autopoiesis.

Conceptualism: See: Conceptual realism.

Conceptual nominalism: The view that concepts, ideas and classes only have an existence in the human mind and in human knowledge categories.

Conceptual realism: The view that concepts, ideas and classes have an existence independent of the human mind and human knowledge categories.

Convergence: The co-influence of systems properties.

Durée/duration: Lived time, as opposed to time as it is measured and quantified.

Ding an sich: See: Noumenon.

Divergence: The individuation of identity from within complex systems.

Emergence: The rising out of complex systems of properties and structures (see: Convergence and divergence).

© Springer Nature Switzerland AG 2021
L. Landeweerd, *Time, Life & Memory*, Library of Ethics and Applied
Philosophy 38, https://doi.org/10.1007/978-3-030-56853-5

Epiphenomenalism: The neurophilosophical view that mental events are caused by physical events, not vice versa (Aldous Huxley adhered to this position, stating that the steam whistle of a train does not contribute to the locomotion of that train either).

Finalism/finality: See: Teleology.

Individuation: The organic process of becoming a self, either psychological or biological (see: Divergence, convergence, emergence and complexity).

Ineffable, the: Beyond description, beyond words, beyond our common ways of comprehending.

Metaphysics: Metaphysics is usually associated with that aspect of philosophy that is preoccupied with the problematic concept of 'being'. The word 'metaphysics' however has a rather mundane origin. In the original canon of Aristotelian philosophy, two important works came to be distinguished: his writings concerning first philosophy ('τὰ περὶ τῆς πρώτης φιλοσοφίας') and his writings on second philosophy, or 'further philosophical writings' ('τὰ περὶ τῆς δευτέρας φιλοσοφίας') that were later dubbed the 'physics' ('τὰ φυσικὰ'). Andronicus of Rhodes who edited Aristotle's works in the first century before Christ is said to have placed the writings on 'first philosophy' on a shelf to the right side of the 'physics', calling them 'τὰ μετὰ τὰ φυσικὰ βιβλία' or 'the books [scrolls] that come after the [scrolls on] physics'. This was misread by Latin scholastics, who thought it meant 'the science of what is beyond the physical'. Scholars in later centuries searched for an intrinsic explanation for Andronicus's term 'metaphysics'. In the ensuing centuries, metaphysics came to hold three meanings: (a) the field that studies the non-physical realm – the esoteric (which has unfortunately resulted in the practice of placing philosophical publications next to books on horoscopes and divination in your average second hand bookstore); (b) the field that studies the world from theological perspectives – which equates philosophy and theology; and (c) the field that researches issues of 'being' – ontological questions about the nature of life, the nature of the human being, the status of knowledge etc. In this book I use metaphysics in this latter meaning.

Negentropy: Going in the opposite direction from entropy; demonstrating properties of selforganisation (see: Autopoiesis); e.g. life.

Nominalism: See: Conceptual nominalism.

Noosphere: A concept coined by Vladimir Vernadsky and further developed by Édouard LeRoy and Pièrre Teilhard de Chardin that describes the sphere of all human thought (sometimes including all knowledge as well as all information) that emerged over the preceding layers of the biosphere (all the sphere of all life) and geosphere (the globe), thus presenting a fundamentally new evolutionary development.

Noumenon/-menal: The world referred to, but at the same time, the world as it retreats from reference.

Phrenology: A pseudoscience involving the measurement of bumps on the skull to predict mental traits.

Phenomenon/-al: What appears to us, the world of experience.

Problem of the Universals: See: Universals.

Realism: See: Conceptual realism.

Selforganisation: See: Autopoiesis.

Substance/substantivism: The philosophical view that the basis of all that is real is to be found in the nature of the costituent substances. Some philosophers regard this in the same light as the 'thingification' of the worldview: the idea that everything real can be reduced to 'the collection of all things', thus negating the concept of change, or even the importance of 'events'. Things however have an individual and discrete nature, whilst substances are amporhic, without form. Substantivism is thus traditionally regarded as the theoretical position that prioritises substance over form, rather than the theoretical position that considers the universe in terms of being the collection of all discrete things.

Technoscience: The increasing merger in contemporary science between science and technology, research and engineering.

Teleology: Teleology, also referred to as finality or finalism is the originally Classic Greek, Aristotelian school of thought that saw all processes as striving for an already implied end state. Thus, something could be explained in terms of the function of its end state, its purpose, or its goal. In mathematics, the concept of an attractor is teleological: it is a set of numerical values, toward which a system tends to evolve in spite of the variety of starting conditions of the system.

Temps/ clock time: Time rendered measurable, either in physics or in daily life.

Thing in itself: See: Noumenon.

Universals (battle of): The battle between Aristotelians and Platonists in the late middle ages over whether classes or kinds (universals) should be regarded as dependent (Aristotelian: conceptual nominalists) of human knowledge or the mind, or as having an independent (Platonic: conceptual realists) existence.

Vitalism: *in biology:* the mostly defunct theory that living organisms are fundamentally distinct from non-living entities, living organics from dead organics and an-organics, due to some non-physical element, spark of life, and as such are governed by different principles than are inanimate things.

Vitalism: *in philosophy:* the philosophical view that the phenomenon of life cannot be accounted for satisfactorily through a reduction to mechanical principles.

Printed in the United States
by Baker & Taylor Publisher Services